The Brighton Women and Science Group came together in 1976. The group was formed as a discussion forum, its members both scientists and non-scientists, its aims to focus attention on the enormous issues raised by scientific research and its applications, and the crucial ways in which science affects women's lives. Members of the group have held conferences on these themes, and have participated in those held by the British Society for Social Responsibility in Science and the American Association for the Advancement of Science.

This book was conceived as a result of the group recognising a serious gap in the literature on science. The authors explore the impact of science and technology on women's lives and society as a whole, their uses in perpetuating social divisions between the sexes, and theories on the innate nature of the sexes based on scientific propositions. The profound effects of science on familiar areas such as contraception, mental health, childbirth, and the less familiar — selective breeding, psychological testing, theories of evolution, are analysed in the light of recent research. Written for the general reader, the scientist and the social scientist alike, *Alice Through the Microscope* confronts issues which radically affect the quality of our lives and the future of our society.

List of Contributors

Pat ALLEN is doing postgraduate research in metazoan genetics at Queen's University, Belfast.

Sandy BEST has worked in Women's Centres and Community Resource Centres.

Lynda BIRKE is a Research Fellow in Biology at the Open University in Milton Keynes. Her research interests include the effects of hormones and behaviour.

Libby CURRAN is a graduate of Sussex University, now doing postgraduate research in education, and particularly intersted in the structure of the classroom and language.

Wendy FAULKNER is a graduate in Biology from Sussex University. She is currently working towards qualifying as a dietician.

Jalna HANMER is a Lecturer in the Department of Applied Social Studies at Bradford University. Her research interests include violence in the family and reproductive technology.

Deirdre JANSON-SMITH is a graduate in Zoology from Edinburgh University, and is doing postgraduate research on the history of women in medicine.

Janet SAYERS is a Lecturer in Psychology at the University of Kent. She is interested in the psychology of sex differences, and Freudian theories as applied to women.

Hilary STANDING is a Lecturer in Social Anthropology at the University of Sussex. Her current research interests include medical anthropology and women's studies.

Ruth WALLSGROVE has a degree in the soft sciences. She works as a member of the Spare Rib collective.

Vivien WALSH works in the Science Policy Research Unit at Sussex University, and is researching into the impact of new technologies on workers.

ALICE THROUGH THE MICROSCOPE

The Power of Science over Women's Lives

by

The Brighton Women & Science Group

Edited by
Lynda Birke · Wendy Faulkner · Sandy Best
Deirdre Janson-Smith · Kathy Overfield

Virago

London

Published by VIRAGO Limited 1980
5 Wardour Street, London W1V 3HE

Copyright © Brighton Women & Science Group 1980

ISBN 0 86068 078 9 Hardback
ISBN 0 86068 079 7 Paperback

Typeset by Malvern Typesetting Services
and printed in Great Britain at
The Anchor Press, Tiptree, Essex

For our mothers, our sisters

Contents

Part One

SCIENCE AND WOMEN IN SOCIETY

Introduction

This book, as its title implies, is about women and science. We wish here to outline why we decided to write it, dealing as it does with a topic that has been rarely addressed until quite recently. The introduction is divided into four sections. First, we deal with the question, why science? What is the relevance of science to women's lives? The second section deals with the question, why women? While there is now a rapidly growing literature on the politics and sociology of science, there is very little to date that has been written specifically about women. We outline why we consider a feminist critique of science to be necessary. The third section considers science as system and mode of thought, and addresses particularly the 'use–abuse' model – that is, the argument that implies that science itself, as a body of 'facts', is ethically neutral, and that it is only the uses to which society puts scientific knowledge that can be 'good' or 'bad'. The fourth section outlines how the book is organised.

In this book we are critical of a number of aspects of science, and in particular we are critical of some of the theories, produced in the name of scientific knowledge, about women. We believe them to be, more often than not, based on sexist prejudices and ideology; we also believe such theories to have a specific role in providing scientific legitimation of the existing hierarchical organisation of our society.

Most people today are critical of some aspects of science, even though they may know or understand little about it. Few, for example, would condone the research that gave us the atom bomb, nerve gases, or widespread pollutants and defoliants. In criticising science we are, therefore, adding our voices to those of many others whose criticisms of the existing form of science have been loud and long.[1] But we must also qualify our criticism. Readers may get to the end of a book such as this, and feel that its contributors have a marked distaste for science and its practice. This is not so. Several of us are scientifically trained, and a few are practising scientists with a lifelong concern with the subject. It is not the scientific method *per se* that we criticise, but the distortion of that method in the service of particular ideologies. Although writing primarily about women, we are equally critical of the kind of science that purports to provide evidence for the 'naturalness' of racial inequalities, or class inequalities, or the idea, for example, that homosexuals are sick. And while the pages of this book deal with notions of how *women's* biology determines their behaviour, we are as critical of similar notions applied to men – the idea, for example, often quoted in the popular press, that men are 'naturally' aggressive owing to some feature of their biology.[2] In short, we do not believe that the existing social order – based as it is on a variety of inequalities and hierarchies – is biologically, naturally, *inevitable*.

I WHY SCIENCE?

Our lives have been greatly influenced by science. The scientific and technological revolutions of the last few hundred years have changed our way of life to such an extent that it would be unrecognisable to our ancestors. In the wake of these rapid changes, people tend to have certain beliefs about science – in particular, that it is a value-free, objective source of knowledge and information. One consequence of this widespread belief is that, if something is said to be scientifically 'proven', it is necessarily assumed to be true. Believers in the truth of scientific proofs, however, do not always ask themselves what is meant by 'proof'. Many people, for example, continue to believe the much publicised claim that black Americans have lower IQs than white Americans, despite several criticisms of the scientific and social validity of this claim.[3]

Science is a field of inquiry that asks questions about the world in which we live, and provides predictions about the way that world works. The knowledge obtained can then be put to use in a variety of ways. At first glance, such inquiry seems fairly neutral. Indeed, many of us have an image of The Scientist working away at some abstract problem that fascinates him (and scientists often are thought of as 'him') in a small laboratory, seemingly far removed from the world. But this is rather an outmoded image. On looking deeper, we find that the form of science is no longer so benign: the majority of scientists today do not work individually, on problems of their own choosing. Most are financed by industry, or by government defence budgets, working on problems geared specifically to the needs of private profit, or to the development of a military–industrial complex. It is a sad reflection on the values of our society that relatively little of the annual research expenditure in any Western, industrialised country goes into medical and welfare research. Considerably more of it (in the region of 25 per cent in the UK) goes on defence.[4]

Science, like anything else, is a social activity, moulded by the society in which it takes place; in order to change it, to use scientific knowledge primarily for *human* benefit, we must not only understand it as a body of facts and theories which can themselves be distorted by human values, but we must also come to understand it as a social activity carried out in ways determined by those values. In a hierarchical society, those values reflect the power of certain groups over others.

There are basically two themes of 'why science?' The first concerns scientific practice, and the specific uses to which results are put within the process of production. It has been pointed out that 'production science (that is, scientific work geared towards industrial production) is science for profit, science for the accumulation of capital . . . it is concerned with developing industrial capacity, exploiting new materials, increasing profitability'.[5] While some of this may in some way benefit people – for example, in the production of new drugs which help in the control of disease – human benefit is not the major objective of the industries employing the majority of scientists: the major objectives are the maximisation of profit and of capital.

It is also worth pointing out that even when a development of some benefit, such as a new drug, occurs, *how* and *when* it gets to the people who need it may be determined by economic and political considerations: many people continue to fall ill or die because the necessary drugs are too expensive, or simply not available.[6]

We are not, however, concerned primarily with this facet of modern science, as critiques of this kind have appeared elsewhere.[7] We are more concerned with a second face of modern science, science in the service of social control. 'Social-control science takes two forms: it is related either to defence against potential external enemies, or the development of techniques for the pacification, manipulation and control of the indigenous population.'[8] This face of science includes the development of more and more sophisticated weaponry for use against people of other nations. It also includes the development of techniques which, we are assured, will help to control otherwise uncontrollable people.[9] Psychosurgery, for example, has been advocated for a number of people defined as deviant[10] – including politically militant people, or those who have preferred not to take part in the merciless slaughter of a war they did not choose to make.[11]

Social-control science affects women in subtle ways. We find scientific definitions implying that we are incapable of certain tasks, that we are biologically destined for domesticity and passivity. These are the themes of much of this book. Some of these definitions are embedded in strictly scientific language, and so are not particularly accessible to anyone wishing to study them. To illustrate our point here, however, we can look at a few examples from nineteenth-century science. The absurdity of these definitions is clearer and more obvious, seen from a distance, and in the light of rather different attitudes toward women now.

Science and women's oppression in the nineteenth century

The role of science in confirming women's oppression is not new. During the nineteenth century, when biologists and theologians were furiously debating the ideas of evolution, there were suggestions that women had not evolved as much as men.[12] Biological and social evolution were equated, and the equation went thus: men (preferably white, middle-class, and heterosexual) were socially superior to women; therefore, women could not have advanced biologically as much as men. This was held to be due to our reproductive functions, which, because of their insatiable demands for energy, drew energy from our brains, hence hindering our intellectual development.[13] Women were considered weak and frail as a consequence of this vital energy needed in the womb, and were treated accordingly – at least as long as they were upper-class. Women of the working class had a different standard applied to them: they were expected to stay healthy enough to carry on working, whether in the mines, factories, domestic service, or in producing a new generation of labour.[14]

The idea of 'vital energy' was given scientific credibility by scientists and doctors alike:

If we wish woman to fulfil the task of motherhood fully she cannot possess a masculine brain. If the feminine qualities were developed to the same degree as those of the male, her maternal organs would suffer, and we would have before us a repulsive and useless hybrid.[15]

There was enormous popular interest in science during the later nineteenth and early twentieth century: Beatrice Webb referred to this in her autobiography (*My Apprenticeship*) as a 'cult of science', maintained in all the popular magazines of the time.[16] Appeals to 'science' were often used to give credibility to a variety of potentially oppressive measures, such as the appeal against further immigration by 'alien races' in the United States,[17] or the campaign against women's suffrage here. Claims of 'scientific evidence' – however weak – seemed to validate the campaigner's arguments. One such anti-suffrage campaigner wrote, in 1913, that women's minds were unsuited to making rational decisions as they build up unreal pictures of the external world. The explanation, he maintained:

is to be found in all the physiological attachments of woman's mind: in the fact that mental images are in her over-intimately linked up with emotional responses . . . that intellectual analysis . . . involves an inhibition of reflex responses which is felt as neural distress . . . and that woman looks upon her mind not as an implement for the pursuit of truth but as an instrument for providing her with creature comforts in the form of agreeable mental images.[18]

Similar 'scientific evidence' was marshalled to oppose women's attempts to obtain higher education: they were unlikely to be able to study owing to the insatiable demands of their reproductive systems. One Alabama doctor warned:

Women beware. You are on the brink of destruction. You have hitherto been engaged in crushing your waists; now you are attempting to cultivate your mind . . . You have been incessantly stimulating your emotions with concerts and operas, with French plays, and French novels; now you are exerting your understanding to learn Greek, and solve propositions in Euclid. Beware!! Science pronounces that the woman who studies is lost.[19]

It was not only in justifying women's social oppression that the medical and scientific professions excelled. During the 1870s there was much scaremongering in the pages of medical journals on the terrible consequences for women (or for men) should they try to practice any form of birth control. For women, the consequences included:

death, or severe illness from acute and chronic metritis, leucorrhoea, menorrhagia, and haematocoele . . . cancer, in an aggravated form assuming in such examples a galloping character so rapid in its course, ovarian dropsy and ovaritus. . . . Lastly, mania leading to suicide and the most repulsive nymphomania are induced.[20]

Most patients (and probably many of our readers) may not have understood the terms. The important thing is that they sound fearsome.

These are a few examples of how scientific arguments and reasons were put forward to legitimate woman's position in society. These arguments supported the view that, if her position is due to biological inferiority, there is little point in trying to change it. It is not enough to dismiss these examples (and the many more examples from the twentieth century in the following pages) as either laughable, or as isolated, deviant incidents. The ways in which such arguments were – and are – used rest on assumptions that 'science' is a body of indisputable facts, untarnished by ideology. It is our central theme that such assumptions are false.

II   WHY WOMEN? WOMEN AND SCIENCE AS IDEOLOGY

In writing about women and science, we are not saying that science has adverse effects only on women, or that the adverse effects of which we write apply to all women equally. Some of the points made by the authors may apply only to women (men do not menstruate or have babies, for example); some may simply apply differently to women and men (as, for example, in the definitions applied in mental illness, or the process whereby women are ousted from scientific practice).

Our intention, as editors of this book, was to highlight *some* of the ways in which scientific changes have affected women and women's oppression in our own society. We have concentrated mainly on women in Western society. This is not to imply that women in other cultures are any less affected than ourselves: we know that this is far from the truth. They may, however, be affected in different ways. It would have been exciting and interesting to tackle our subject on a cross-cultural basis – and it would have kept us occupied for years; so we decided to stay at home.

But back to the original question: why a book on science and *women?* First, because there is now a considerable literature on the politics and sociology of science, but very little from a feminist perspective.[21] Second, we are feminists, and many of us are scientists: we are therefore interested in the ways in which one sphere of our experience influences another. Third, despite considerable interest shown recently in women's studies, resulting from the Women's Liberation Movement, little attention has been paid to science. Women's studies still tend to focus on areas traditionally considered more suitable for women (such as sociology, or literature studies), and have tended to ignore traditionally 'male' subjects, such as science or engineering. Science, it is true, does not belong to us, but that is no reason to ignore it.

> In fact, the title 'science' has been exclusively reserved for that knowledge and those skills which can be systematised and incorporated into the academic culture of the ruling capitalist class. All other knowledge and skills that belonged to the popular culture, and which have accumulated over centuries of careful and selective observations and practice, have been denigrated and labelled unscientific.[22]

That the skills of the 'popular culture', including women's skills, have been systematically denigrated is bad enough. That feminism colludes in this by ignoring that from which we are excluded is worse still.

Missing from the scene

If you enter a science laboratory in any university or industry, you will find that there *are* some women about. They are usually employed as cleaners, secretaries or laboratory technicians. They are all essential to the smooth running of a research laboratory, but they rarely participate in important decisions related to the research programme. In other words, women are predominantly found in supportive, supplementary, roles. The highly prestigious roles – be they research scientists or senior administrators – are filled mainly by men.

If science is seen as a rational and objective pursuit, then this is a social judgement which also tends to exclude women from its practice, for these are the qualities that are defined by social stereotypes as 'masculine'. To be 'feminine' in our society is too often to be thought irrational, subjective, emotional – and unfitted for responsible jobs. Thus our very 'nature' disqualifies us from gaining any say in an area that nevertheless daily increases its control over our lives.

To be excluded from science as a woman is to be powerless in a specific way. We are not missing from the scene, yet we are missing from the centre stage. The picture within the scientific world is no different from that within any other sphere of society: the actors change, the play remains the same. Women are alienated from the activity called science even while we work within it. The picture of science as some kind of anonymous and autonomous 'thing', inaccessible to women, becomes not merely an image but a reality – it is alienating not because science stands apart, but because it reflects the deep sexual divisions of our society in their most powerful and insidious forms.

Is biology our destiny?

As women are so often defined and limited by their biology, this book is necessarily based largely on the biological and psychological sciences.[23] The biologist who turns from a study of animals to a study of human beings inevitably brings with her/him the social values of their time. The theories that s/he then develops about human behaviour are very likely to be influenced by those values, whether or not they are acknowledged. Some (male) scientists are avowedly chauvinistic; one can see this in some of the simplified versions of their findings for the popular press. What we often find publicised are the areas of biology that support the existing social system. We find articles indicating that the 'latest scientific findings' show that women are 'naturally' nurturant, while men are 'naturally' aggressive. Claiming that these are scientific truths serves to lull the reader into believing them without question.

In writing up their results, and interpreting them for publication in the scientific press, scientists have sometimes implied an 'innate' inferiority based on sex, race, class or sexual preference. These implications do not always lead to clear predictions, let alone ones that can be tested experimentally. Their chief function is ideological. The current popularisation of sociobiology[24] is just such an example. Popular versions 'explain' male philandery and female faithfulness: they 'explain' Man as the go-getter and Woman as the home-lover. Yet the evidence for such a view is, to say the least, tenuous (see Chapter 3). If such explanations of the current social order, in terms of our biology, are to be believed, there would be little point in our trying to build a more egalitarian society, and little hope for more respectful relationships between human beings. We are, if these popularisers of science are to be believed, wasting our time.

Notions of biological destiny have also invaded our lives as mothers. About thirty years ago there were many reports in the scientific press claiming that a child's psychology is permanently damaged if her/his mother is not with her/him continually. Separation, for however brief a period, was held to be deleterious: a mother should stay where the expanding economy of the time needed her most – at home. Her role as consumer at this time was second only to her role as mother. Those who disagreed with this view were, in general, ignored.

The suffering allegedly caused to the child by a working mother was called 'maternal deprivation',[25] and was given credibility by appeals to 'scientific proof'. Not to feel the maternal instincts – to want to get away from the demands of children for a while – was held to be unnatural. How many women must have felt – and still do feel – tremendous guilt, simply because medical and scientific experts have assured her that she must become the centre of her child's universe if that child is not to grow up a delinquent?

The notion of 'maternal deprivation' has been criticised more recently.[26] It is undoubtedly true that children need security; they do suffer if the normal tenor of their lives is disrupted. But, as more recent studies have shown,[27] the constancy can be provided by the attention and love of another individual or individuals: it does not have to be only the biological mother. While science is less inclined now to justifying ideas of maternal deprivation, it still bolsters the idea that women – all women – have 'instincts' that admirably fit them to childbearing and -rearing and to few other interests. However it is phrased, those women who opt not to produce children, or those men who opt to rear their children, are still seen as freaks of nature, as oddities.[28] Biology, it seems, is meant to be our destiny.

The 'maternal deprivation' myth, like so many others concerning women, has been sold to us as the only rational truth, as serving the interests of humanity. Deprived children, it was claimed, would never be useful members of society. Thus women were persuaded that it was for the common good that they 'sacrificed' themselves to motherhood. Even now that the link between maternal absences and later delinquency has been discredited,[29] women's guilt remains. Many women remain anxious that they will not do the best for their child if they work outside the home. This is an anxiety that is perpetuated by the way in which mothers are portrayed in radio and television programmes dealing with parenthood, which tend to locate the responsibility for the child firmly with the mother.[30]

The invasion of technology

In this scientific age, it is predominantly scientific knowledge that shapes attitudes, rather than the knowledge gained by, say, insight or meditation. But the technological products of science also bite deep into our lives, whether as workers outside the home or workers within it.

We tend to think of our technological gadgets in the home as liberating, since they apparently save labour. But is this really so? It has been pointed out that most of the domestic inventions that we now take for granted – the vacuum cleaner, the washing machine and so on – came on to the domestic scene accompanied by a wave of publicity encouraging women to do far more cleaning than had been done previously.[31] At the end of the eighteenth century, women were involved in many more tasks than simply cleaning: '[they] weren't just making apple pies and embroidered samplers; they were making bread, butter, cloth, clothing, soap, candles, medicine, and other things essential for their families' survival'.[32] Apparently they had little time left for cleaning. Yet during the nineteenth century, things had 'drastically changed: the traditional home crafts were vanishing into the factories. . . . Cloth, and soon candles, soap and butter, joined buttons and needles as things that most women *bought* rather than made' (emphasis in original).[33]

This left a lot of time to be filled by the housewife, and by the end of the nineteenth century there was a drive towards Hygienic Housework. Not only did this have to be *scientifically* based, but it also had to be done at frequent intervals – if possible, at least once a day. Articles appeared in women's magazines expressing horror at the dangers of household dirt: a single speck of dust carried untold dangers.[34] It was into this environment that all the new technological devices came – an environment in which women were being encouraged to clean every available surface in a daily frenzy. Perhaps, then, the advent of 'labour-saving devices' has not been as liberating as we tend to think.

The new justification for housework came in terms of scientific explanations: women were told that they should learn basic science in order to practice 'scientific housekeeping'.[35] In this way, a form of science education was developed for girls which was different from that for boys – domestic science. This was of fundamental importance:

> when the grand meaning and hidden power of her ordained sphere dawn
> upon her in their full force thru [sic] scientific study, then she will not

sigh because Nature has assigned her special duties which man has deemed safe to be trusted to her instincts, yet in reality need for their performance the highest scientific knowledge.[36]

The development of this special kind of science education has contributed to the idea that *ordinary* science is more suitable for boys than for girls (see Chapter 1).[37]

An important feature of this special science education for girls was that it was intended to provide a knowledge of basic science relevant to the home – basic hygiene, chemistry and so on. It did not, however, include a training in how to mend the domestic machines when they went wrong. Things are little different today, and few women feel themselves to be capable of understanding or mending even simple technological devices: 'I don't understand it, but I'll find a man to mend it' is all too common a response.

If women in general continue to accept powerlessness in this way, we are unlikely to learn how science and technology can contribute either to our oppression or to our liberation. We need to understand how biological theories contribute to our position in society, and how supposedly liberating technological advances can in fact catch us in a trap. Not only do we need that knowledge to increase our understanding of the role of particular ideologies within modern science; but we also need it so that we might build a better, and more humane, society.

III SCIENCE AS SYSTEM AND MODE OF THOUGHT

It is difficult to avoid talking or writing about science as though it were some kind of monolithic 'thing', with a life and authority of its own, answerable to no one. Yet to understand the role of scientific knowledge in our oppression we clearly need to avoid this popular image of science as 'purely' a body of knowledge that somehow unfolds under its own momentum, leaving a trail of useful facts as it rolls on.

To attempt to do this, we want first to point out a few things about scientific method, since not all our readers will necessarily be familiar with it. In so doing, we acknowledge that we are giving a very brief outline, which to scientifically trained readers may seem rather crude. We apologise for this, but feel it to be necessary for the benefit of those who have not worked within science. We also want to examine the image of science offered in a particular model of science in society, the 'use–abuse' model.

Scientific method

Scientific method seeks to gather information about the natural world by observation, by experimentation, and by reasoned argument, rather than, say, by meditation and philosophical discussion. We might sum up the method as follows. The first step in most scientific investigations is to set up a hypothesis of how something is. This hypothesis is deduced from a knowledge of previous scientific investigations which have provided certain facts about the natural world. Once a hypothesis is established, it is used as a basis from which to make certain predictions – and these predictions should be ones that can be tested experimentally. The experiments are then designed and carried out, and the results compared with the original predictions. If they agree, the hypothesis stands. If they do not, then, ideally, the hypothesis is rejected. In practice, of course, scientists being only human, they may sometimes reject the apparently errant results, and assume that they have measured something incorrectly, rather than immediately rejecting the hypothesis. If the data agree, further testing of the hypothesis, using the same or different experiments, might be carried out. If further testing of the hypothesis is also successful, the hypothesis attains the status of theory, or of law (e.g., Darwin's theory of natural selection, or Newton's law of thermodynamics) – when it is considered, for practical purposes, to need no further direct questioning, at least for the time being.

Theories may persist for years, even centuries. Any data that are in apparent contradiction to the major theory may be thought of during that period as simply wrong or as exceptional. Alternatively, the theory may be modified or extended to include the errant data. Eventually, however, enough 'errant' data may accumulate to throw the theory (and its advocates) into serious doubt, eventually to be replaced by contending theories which seem better able to fit the available data.[38]

The image of the 'objectivity' of science rests on the soundness of the logical argument from predictions to result. The weak link in the chain is often to be found right at the beginning, when the original hypothesis is set up: it is here that the process is most vulnerable to preconditions and assumptions. The interpretation of the data may then be determined by the assumptions on which the hypothesis rests. To illustrate this, we can consider two simplified examples from chapters of this book. For instance, if you believe that women 'naturally' mother, you may set up a hypothesis that predicts that girls should tend to play with dolls. If you then find this to be the case, you may assume that your data support your hypothesis that girls 'naturally' show mothering behaviour. Because this supports your beliefs about what is natural for women or girls, it may not occur to you that the little girls in your experiment might be playing with dolls because their parents encouraged them to do so. Similarly, if you find that lesbians are more self-confident and assertive than heterosexual women, and if you believe that self-confidence and assertiveness are necessarily masculine traits, then you will be likely to conclude that lesbians are in some way biologically masculine. In theory, it would be equally valid to conclude that perhaps all women have these traits, and that somehow these traits have become repressed within heterosexuality. But if you have those beliefs outlined above, this possibility will not occur to you. The closer science gets to human experience, the stronger, it seems, is the possibility of underlying bias to theories.

What we now know as science, and as scientific method, arose at a particular point in the history of a particular society. Science as a way of looking at the world arose at a time when the authority of the church was being challenged, during the seventeenth and eighteenth centuries. Ironically, perhaps, the church itself contributed to the early development of science, in so far as the so-called 'natural theologians' endeavoured to understand the laws of nature, in the belief that they were thereby revealing the perfection of God's design of the natural world. (An example is William Paley (1743–1805), whose book *Natural Theology* was greatly admired by Charles Darwin.)

Scientific method and thought began to offer an alternative to the dogmatic authority of the established church, and to the mysticism of alchemy,[39] which had provided the major explanations of natural phenomena until the sixteenth and seventeenth centuries. 'Science' began to explain the world in a way that was not static, that could not be reduced to superstition or to 'God's Will'. Scientists were to become the high priests of the new religion.

But it was under the impetus of economic and productive expansion during the Industrial Revolution that science began to change: no longer was it sufficient for 'gentleman scientists' simply to describe nature. The material needs of expanding industry were such that science became directed towards solving problems related to the production process, such as finding new materials, finding new ways to use old ones, or automating existing machinery.

Scientific knowledge offered people (at least, those in power) the ability to control the world more than ever before, with consequent possibilities of increasing both wealth and exploitation of the earth's resources. And as industrial capitalism emerged during the nineteenth century, the growing scientific knowledge was used to maintain and justify that system.

Science grew rapidly, especially after the middle of the nineteenth century, and became increasingly institutionalised. No longer a gentle pursuit of individuals, it was fast becoming the concern of large institutions and corporate interests.[40] This served to facilitate an attitude already prevalent in the West – that the earth and her produce are there for our use, for our exploitation, to extend our dominion over the world.[41] Other cultures of the world have not necessarily had this attitude: many have had a more gentle, harmonious relationship with the earth and with nature – one where they are part of nature, not apart from it. North American Indian cultures had this view of the world, and could not understand the plunder and murder of their country and their peoples by white colonisers (who justified their plunder on the grounds that only they represented 'civilised' humanity, the pinnacle of human evolution – the Indian represented a lower stage of existence[42]). One old Wintu woman, witnessing the advent of hydraulic mining in nineteenth-century California, said:

> The White people never cared for land or deer or bear. . . . When we burn grass for grasshoppers, we don't ruin things. We shake down acorns and pinenuts. We don't chop down the trees. . . . But the White people plough up the land, pull down the trees, kill everything. The tree said, 'Don't, I am sore! Don't hurt me!' But they chop it down and cut it up. . . . The Indians never hurt anything, but the White people destroy all. . . . How can the spirit of the earth like the White man? . . . Everywhere the White man has touched it, it is sore.[43]

Perhaps we are justified in referring to the 'rape of the earth'. Western technology seems to allow no space for harmony, either with the earth, or with nature, or with people of other lands. Repeatedly, daily, we rape the earth in order to extract forcibly as much as possible from it in the shortest possible time.

Clearly, some scientific research has given us enormous benefits – better means of combating disease, better devices for the disabled, faster communications and transport and so on. But the same process of scientific development, within the same society, has given us much else besides.

If science is objective, and its practice the 'disinterested pursuit of truth', then scientists, and society generally, are faced with the thorny problem of how science has managed to get its fingers well and truly dirtied. Science in the service of capitalist production and imperialist expansion has given us the atom bomb, the neutron bomb, nuclear reactors, and such niceties as mustard gas and napalm. In the service of social control science has offered us theories of 'innate' differences between people, and has had a role in the development of sophisticated methods of torture, and of brain manipulation.[44]

How is it that the quintessentially 'ideologically neutral' pursuit has got itself into such deep and dubious waters?

The 'use–abuse' model

The explanation often favoured by scientists (and others) is that science offers only facts to the world. What a particular society does with those facts is, according to this argument, outside the control and competence of the scientists, who therefore remain outside blame. Society may choose to use scientific knowledge – perhaps in developing better hearing aids; or it may abuse such knowledge – such as by establishing restrictive education programmes for black children, on the grounds of supposed innate differences in intelligence. The pursuit of truth, it is often claimed, is above such things. Indeed, it is positively harmful to try to impede such a noble task.

The 'use–abuse' model serves two purposes. First, it absolves scientists themselves from responsibility, and from the need to question the ethics of what they are doing. More important, it bolsters the image of 'pure' science, somehow divorced from the social world in which it takes place, and thus diverts attention from more fundamental questions of the politics of science. Those who hold these views tend to create the impression that ideas occur to scientists solely as a result of detached, objective, thought. Perhaps scientists approached this ideal when thinking about, say, the structure of the atom. But it is much less likely when they are thinking about their own species. Scientists set up their hypotheses about human behaviour, having spent many years as a part of that human society. The 'facts' that then emerge from investigations are likely to be influenced by this social experience.

The 'use–abuse' model, however, sets up another dilemma for the 'socially aware' scientist who attempts to come to grips with these problems. If scientists are producing knowledge that is potentially dangerous, or threatening to the social order, what should they do about it? Within a model that hails the objectivity of scientists, the only possible answer seems to be that those who produce the knowledge should also control its use. But this would only serve to increase the elitist nature of science, removing the control of policy decisions about science even further from ordinary people.

What the 'use–abuse' model ignores is the social reality of the activity called science. Scientists do not work in a social vacuum: they have to compete on a (highly competitive) job market; they have to obtain funding from various organisations for their research (if, that is, they are free to have any say in the direction of their research at all: many are not, owing to constraints imposed by the needs of the organisation employing them). Obtaining funding itself often imposes limits on what is researched – the problems dealt with, the interpretations given, and the subsequent uses of the research. An instance of this is referred to in Chapter 9, in which it is pointed out how the problems of obtaining funding severely curtailed early research into sex and sex hormones. It was not until the more 'progressive' social environment of the postwar years that the research programmes really made much progress.

Rather than serving humanity, then, much of science is geared to the needs of production within industry. Frequently, of course, the science being carried out is related not to real needs (such as improving the distribution of food throughout the world), but to creating false 'needs' within this society. At its most absurd, we have such things as industries that produce better budgie diapers so that little Polly can fly around the house without leaving a trail of euphemisms. And this in a world in which half the population do not have even the basic necessities of life. Some scientists may be trying to see their way through to social responsibility, but they will not get far within a model that ignores the social nature of science.

We have gone into the use–abuse debate for two reasons. First, it is necessary to make clear our rejection of it because it *is* so pervasive, and persuasive (most people believe, to a greater or lesser degree, in the objectivity of science, and correspondingly trust in the technocrat). Second, only when it is examined does it become possible to ask questions about the potential nature of science in a feminist/socialist framework. Once we realise that 'science' is, and must be, shaped by society, rather than existing in a vacuum, then we may hope that science can be made to serve human needs, rather than the needs of profit. This conjures up images of 'state control' and 'gross manipulation' to many people – the idea that the direction of scientific research could deliberately be altered for social benefit seems abhorrent to them. To us this seems a naïve response. Not only do we have manipulation of scientific research already, but we would hope that, in a truly humane society, manipulation for human benefit might be considered to be a good thing. [45]

As feminists, as socialists, *and* as scientists, we believe that it is vital for women to come to understand both the processes and the function of science within patriarchy. It is for this reason that we wish to begin a feminist analysis of science with this book.

IV HOW THE BOOK IS ORGANISED

Part I primarily discusses the literature on differences or similarities between women and men. The first chapter is about how accessible science is to women. We have said that science is essentially mystifying to the majority of women, and that this mystification is important if women are to be denied control over scientific progress. The alienation of women from science makes the field inaccessible, and this begins in the educational system. Although there have been slight improvements in the last two decades, science teaching still reflects the fact that it is considered more suitable for boys: even the examples still used in many textbooks are irrelevant to many girls. The first chapter therefore concentrates on science education.

The next two chapters look at two aspects of scientific theory that have something to say about the roles of women and men. Both the literature on the psychology of sex differences and that on sociobiology carry examples that claim to have proved that women are 'naturally' inferior in a number of ways. It is, of course, perfectly valid (in scientific terms) to ask why women and men are different; why, and in what ways, they behave differently in the world. What is less valid is that the questions asked are almost always those that will provide an answer supporting the conventional views. The questions asked are determined more by the prevailing ideology than by 'scientific rationality'.

Part II is devoted to examining specific examples of some ways in which science has contributed to definitions of women. We can see from these chapters that much of the scientific work done in this field works on the same assumptions as those held by society at large. Thus we see, in the chapter on mental health (Chapter 6), that our society tends to view women as inherently more sickly than men – a view that the literature tends to mirror. Similarly, we find that women's sexuality, and their reproductive functions such as menstruation, tend to be oversimplified and misunderstood.

Part III deals with the relation between science, technology and women's powerlessness over aspects of their own lives, and especially their reproduction. Technology surrounds us even in the act of giving birth, and robs us of the power to control even that act. Chapter 9 deals with the historical development of the technology which gave us the Pill, and illustrates how easily a technology that is of potential benefit to millions of people[46] can be held up, and shaped, by political and economic considerations. Chapter 10 deals with what at first seems like science fiction, but has immense implications for women: reproductive engineering. We have already seen the birth of a baby conceived outside her mother's womb, and the promised possibility of technology enabling us to choose the sex of our children. Some of the implications of these developments for women are quite horrific. The final chapter in the book deals with some speculations on the role, and philosophy, of science, and the implications of these for feminists.

Much of the material covered in this book is aimed at *beginning* an analysis of how science can influence the lives of women. It is only a beginning, and necessarily rather sketchy. Much of it seems pessimistic, even cynical. We cannot apologise for that, although we hope that it stimulates thought, discussion – and above all, action to change the situation.

The book has been edited by the Brighton Women and Science Group, which was formed in the summer of 1976. It was set up with the aim of finding out the extent to which scientific findings affect women. Some we knew to have obvious implications for women – such as the research that led to the Pill. Much of science we soon discovered to have indirect, less obvious, implications, which we hope you will discover in reading the book. We are aware that, not only has there been virtually nothing written about women and science[47] (at least, little that is reasonably accessible), but that many women thought it was irrelevant to their position as women. For these reasons we felt that a book on the subject was overdue. As far as possible, we have tried to avoid the use of mystifying, scientific, academic jargon, as well as political polemic, so that the book will be more accessible to those who do not have much understanding of science (or of the Women's Liberation Movement). We hope that we have succeeded.

As editors, we are responsible for the final form of the book – but not necessarily for the content of all the chapters. We do not necessarily agree with all of the views put forward by the authors.

Nor do we see the different contributions as definitive: the book is an exploration, a beginning, in an area that is of vital concern to women.

The book has been edited by Sandy Best, Lynda Birke, Wendy Faulkner, Deirdre Janson-Smith and Kathy Overfield, with the assistance of Libby Curran. Other members of the group who have contributed to the ideas contained in it are Lynn Benjamin and Charmian Kenner. Col Eglington also contributed useful ideas in discussion while she was in Britain.

One final note. Scientific information changes, sometimes quickly and significantly. We have tried to make this information as up to date as possible, but inevitably there will be some material that becomes out of date between the writing of the book and its publication. For this, we apologise.

1

Science Education:
Did She Drop Out
Or Was She Pushed?

Libby Curran

SOME FACTS AND FIGURES

Girls do slightly better than boys in primary school, according to achievement tests at eight and eleven years, but after that it's downhill all the way . . .

Only slightly more boys than girls take O levels, but nearly twice as many go to university. By postgraduate level there are nearly three times as many men. In the last few years male demand for university places has fallen, probably because of graduate unemployment: a university degree is no longer a guarantee of employment, and more boys are opting for job training. But in 1971, for instance, there were five times as many men as women at postgraduate level.

The higher up the ladder of educational success, the fewer females you will find. These biases are greatly exaggerated within science

Table 1:[1] *Number of men for every woman at different stages of the educational system: for all subjects, science subjects, and engineering and technology (December 1975, Great Britain)*

	All subjects	Science	Engineering and technology
A level	1.3	2.8	—
Undergraduate	1.8	2.4	24.0
Postgraduate	2.8	5.0	19.5

subjects. Women tend to study the biological sciences, and men the physical sciences. Engineering and technology is an almost exclusively male field.

Beyond the university, this pattern is repeated in the job structure: the higher up the ladder of occupational success, the fewer women there are. If a job has high status, in terms of money, power and prestige, you can be sure it will be a job done predominantly by men. Most primary schoolteachers, for example, are women, but most primary school heads are men (75 per cent). So are most university lecturers and virtually all professors (see Table 2).

Although women are 10 per cent of persons with scientific and technological qualifications, they make up less than 2 per cent of managers, practising research scientists and technologists. The 1966 Sample Census (a one-in-ten sample) shows that there are eleven male company directors with science backgrounds, and not a single woman.[3] There are 200 male managers for every woman. The majority of women with scientific training end up unemployed, or teaching (in schools), or doing low-level jobs – working as technicians, lab assistants, computer programmers, generally helping male scientists with their work. As usual, women are 'servicing' men (see Table 3).

Table 2:[2] Number of men for every woman, full-time teaching and research staff (1975, Great Britain)

	All subjects	Science	Engineering and technology
Professors	48.0	113.4	No women
Readers, senior lecturers	15.0	25.8	108.6
Lecturers and assistants	6.5	12.0	28.7
Others	2.5	3.9	7.2
Total	7.7	13.0	31.7

When only a small percentage of girls take science at school, and only a small percentage of these go on to university, and only a small percentage of women graduates end up in key positions, we are left

'Men outnumber women by five to one at postgraduate level in science.'

Table 3:[4] *Occupations of men and women with scientific or technological qualifications**

	Percentage of qualified men	Number	Percentage of qualified women	Number
		(One in ten)		
Managers	12.0	3,451	0.5	17
University teachers	3.6	1,034	2.1	71
Teachers	13.4	3,832	36.0	1,225
Scientists	12.0	3,442	6.9	235
Engineering and technology	33.8	9,699	0.8	27
Low-level science jobs	6.3	1,801	4.9	165
Others	18.4	5,278	13.7	465
Not in employment	0.5	146	35.1	1,194
Total	100.0	28,683	100.0	3,399

*Numbers exclude those retired or still studying

with a very small number indeed. The number of women involved in developing new technology and advancing scientific knowledge is negligible. Virtually all the positions of power are held by men.

The academic world involves only a small proportion of schoolchildren. Only 7.1 per cent of school-leavers went on to university in 1975–6.[5] What opportunities are open to the rest? The majority of schoolchildren take CSE examinations. Opportunities for further training include apprenticeships to skilled trades; on-the-job training through day-release to further education (FE) colleges; and full-time and part-time FE courses. After three years, there are TOPS (Training Opportunities Scheme) retraining courses.

Generally speaking, the more practical or technical a subject or a course, and the closer it is to training for a practical career, the fewer women you will find.

CSE

At the lower level, even fewer girls study science: there are, for example, 7.5 boys for every girl taking CSE physics, while at O level the ratio is 3.7 to 1.

Apprenticeships

Very few girls enter apprenticeships, partly because they lack the prerequisite skills and qualifications in technical subjects. When there are 88.6 boys taking CSE in technical drawing for every girl, it is hardly surprising that only 8.6 per cent of apprentice draughts*men* are women.[6] In the manufacturing industries, women take up 2.1 per cent of apprenticeships, and in construction 0.3 per cent of apprenticeships are given to girls.[7]

Day release

In 1973 nearly four times as many boys as girls entered day release to study for diplomas in industrial or professional studies[8] (Ordinary and Higher National Diplomas and Certificates). According to 1975 statistics of education, women are under one in five of day-release and sandwich course students.

Further education

Although more girls than boys enter some sort of further education, girls are concentrated in evening and part-time courses; in nursing, colleges of education and lower level, short-term courses leading on to low-status jobs such as typing and catering. Again, women are found in 'service' occupations rather than in 'productive' work. In full-time courses leading to recognised qualifications, men out-number women. This is hardly surprising, since FE colleges are orientated towards engineering and technological courses as training for careers in industry. About one-third of advanced, and one-quarter of non-advanced, courses are of this type (see Table 4).

Table 4:[9] *Number of men for every woman in courses leading to recognised qualifications (December 1975, Great Britain)*

	All subjects	Science	Engineering and technology
FE advanced	3.0	3.5	49.0
FE non-advanced	2.0	1.7	53.4

One bright note: in 1969, only 397 women were taking advanced engineering and technology courses; in 1975 there were three times as many (1,200).

TOPS retraining courses

In craft and technical skills centres in 1977 there were between 23,000 and 24,000 men and 616 women.[10]

So, all in all, girls enjoy the benefit of much less financial investment put into their education; and most end up in dead-end jobs with few prospects.

Girls don't usually do science at school or work in scientific or technical jobs. Why not? There are three possible explanations: because they can't; because they aren't allowed to; or because they don't want to.

BECAUSE THEY CAN'T

The first argument – that girls can't do science – is based on the belief that there are inborn (innate) biological differences between females and males which determine their mental abilities. This may be expressed crudely, as in the assertion that women are illogical, and incapable of abstract thought – a view widely held, even today. Or it may be backed by 'scientific' research.

Some psychologists claim that science attainment is linked to spatial ability: the ability to manipulate three-dimensional objects mentally. Psychological tests have been devised to measure this ability. The ones most commonly used as evidence of sex differences are the Embedded Figures Test, in which the subject is required to pick out a three-dimensional figure from a complex background, and the Rod-and-Frame Test, in which the subject is required to adjust a rod to vertical within a tilted frame. It is a commonly held belief that males perform better than females in these tests, and that therefore the ability is sex-linked and biologically based.

Hugh Fairweather, in an article on sex differences,[11] quotes these remarks from a reference work on spatial ability:

'. . . the well-known fact that girls are inferior to boys in visualizing ability.'

'. . . men doing better than women as nearly always occurs with spatial tests.'

He points out that these statements are unreferenced, and goes on to explode the myth of innate male superiority in this field. In fact, there is no evidence of sex differences in spatial ability in childhood. Studies of young adults show a slight male superiority in a few definitive spatial skills. Since the differences do not appear until late childhood, it is quite possible that they are culturally, rather than biologically, induced. Girls and boys have received different treatment both at home and at school according to their gender.

From the evidence of slight sex differences in a limited number of spatial skills, we cannot jump to the conclusion that females are innately inferior in scientific ability. Spatial ability may be linked to science ability but the two are not synonymous. There are many other factors involved, about some of which there is consensus of agreement, and about others of which there is a great deal of controversy. For example, convergent thinking, logic, independence, self-assertion and conservatism have been variously linked with the scientific mode of thought. It is far from clear if all such qualities are innate and biologically based.

It is important to realise that we are dealing with innate ability and that many factors intervene between ability and achievement. In the nature v. nurture debate, the issue is the relative weight given to innate and the environment of the home, school and society. No one can deny that achievement is a result of the interaction of many complex factors. To attempt to isolate a single variable (in this case spatial ability) is a hazardous exercise.

Whatever the cause of the differences, there is no reason at all why they should affect attainment. Boys display less verbal ability than girls according to psychological tests up to eleven years, but they still go on to do better than girls at every further stage of this verbally orientated educational process.

One last point: teachers who hold the view that children fail at school because of their own inadequacy have a very damaging effect on the children's self-esteem and consequent performance at school. Explanations of children's inadequacy may be biologically or culturally based. For example: working-class children are intellectually inferior by birth; or, working-class children are socially deprived and therefore linguistically inferior by background. Like the theory that girls can't do science because they have less innate aptitude, these theories place the blame for failure at school on the child, rather than on the school or society. Sociological studies [12] have shown that teacher expectations can be self-fulfilling. If a teacher expects a child to do badly for whatever reason (race, class, colour, sex), the child is likely to fulfil those expectations. Without denying individual differences, the teacher should take a positive view of children's capabilities and see them in terms of unfulfilled potential which the teacher must draw out and develop.

If future studies prove conclusively that science ability is linked to spatial ability and that girls are slightly less proficient in this respect, we might conclude that girls therefore need more help in science and practise positive discrimination in terms of encouragement and resources for teaching girls.

So we see there is no evidence for the assumption that girls can't do science, and we must look elsewhere for the answer to our question. Until quite recently the answer was simple: girls didn't do science because they weren't given the chance.

BECAUSE THEY AREN'T ALLOWED TO

The purpose of education, according to the 1906 Board of Education, should be to make children 'efficient members of the class to which they will belong'.[13] The elementary school curriculum should prepare boys for entering 'the lower ranks of industry and commerce' and the girls for becoming good housewives and servants.

To this end, boys did science, extra maths, technical drawing and handicrafts (wood- and metalwork), while the girls did cookery and needlework. Maths, physics and chemistry were considered unsuitable subjects for girls, partly because they were masculine subjects – too difficult for the feminine intellect to understand – but mainly because they would be of little use to the girls in later life. (Basic, scientific principles could be taught through an applied science of the household.)

The teaching of science in the elementary schools was practical, and related to the experiences and 'interests' of the children. It was also instrumental, in that it was to prepare children for future employment. The working-class boy, taught science through handicrafts, is getting practical experience for future apprenticeship. The girl who is taught science through housework is preparing for her job as domestic servant – paid or unpaid.

Meanwhile, the teaching of science in the secondary (grammar and public) schools was very different, at least for boys. The middle/upper-class boy received an education in pure, abstract science, which was supposed to train his mind. The Regulations for Secondary Schools, 1909, suggested that even for older, more academic girls, domestic subjects could be substituted 'partially or wholly for science and for mathematics other than arithmetic'. [14] The different treatment of boys and girls cuts across economic class barriers. In this respect, girls are treated as a separate 'class' from boys, regardless of their economic status.

> The two types of school prepare for the different walks of life – the one for the lower ranks of industry and commerce, the other for the higher ranks of industry and the liberal professions. Consequently, the higher elementary school can only afford to teach a limited number of subjects and with a practical bias, the secondary school has time for more subjects and a more theoretical and academic way of teaching. [15]

Things haven't changed much. It's interesting to compare this statement, written seventy years ago, with the more recent comment:

> What we want is the best education for all . . . whether future leaders of industry, commerce or the professions, or future managers, or skilled or less skilled workers, or housewives and mothers. . . . It should not be a question of success or failure, but on the one hand an academic type of education . . . and on the other hand a more technical, practical kind of education. [16]

The Newsom Report goes on to suggest that average and below-average children should receive an education that is 'practical, realistic, vocational' (p.114) and that girls 'should be educated in terms of their main social function – which is to make for themselves, their children and their husbands, a secure and suitable home and to be mothers'.[17]

More recently, G. H. Bantock argued that the basis of the curriculum for the average child should be more concrete and practical and involve the emotions as well as the intellect. Both boys and girls should receive practical domestic training, 'but perhaps there should be a greater emphasis for girls. The other area of concern would be the technical, which would exploit especially the boys' interest in some of the mechanical inventions by which we are surrounded.'[18]

Children are no longer separated into different schools on the basis of their class backgrounds and future position in society, but comprehensive schools continue to provide different curricula for different 'types' of children, these 'types' fitting conveniently with the roles children will have to fulfil in society.

Academic science for the examination candidates, the future (male) scientists; domestic science for the girls; technical subjects (practical science, woodwork, metalwork) for the future technicians and factory workers.

Girls are no longer completely excluded from science education as a matter of principle, but all too often domestic subjects are considered to be the most suitable preparation for girls' assigned roles of housewife and mother.

Science subjects are supposedly open to girls in our present-day schools, but girls don't usually take them. Sometimes they are not allowed to (discrimination – they are pushed out), sometimes they just don't want to (self-selection – they opt out). Both factors operate at all stages of the educational process.

Primary school

In the primary school, girls manage to do better than boys. What a pity, then, that primary schoolteachers reject science or think it is unimportant.[19] I wonder if this could have anything to do with the fact that most primary schoolteachers are women, and are unlikely to have had a science training at school or college?

There may be open discrimination at the level of organisation and curriculum. Many primary schools still timetable girls to do needlework while boys do a variety of crafts using a range of tools, materials and skills, all standing them in very good stead for later scientific work. Teachers often label activities and toys 'for boys' and 'for girls', and girls may not be allowed to play with mechanical, scientific or constructional toys.

But even if there is no open discrimination, little girls soon learn that science is a boys' subject. If it is taught at all, it is likely to be by a man. Although 36 per cent of women with science qualifications become teachers, as compared with 13.4 per cent of men, the actual number of males is so much greater than females that there are three times as many male science teachers as female. School textbooks reinforce role-stereotyping by showing girls and women in passive, subordinate and limited roles: boys may grow up to be scientists and engineers; girls will become typists, nurses and housewives. Science books typically show male scientists, boys doing all the experiments, and use the pronoun 'he' throughout. Recent books have made an effort to include a few token blacks and females in their illustrations, but most schools cannot afford to restock their entire range. Girls quickly learn that science has little to do with their present or future lives.

It is not enough merely to avoid discriminating on the basis of sex. Boys and girls learn, through the family, mass media and peer group pressures, which activities are appropriate for their sex. Boys are more likely to play with mechanical and scientific toys and to gain practical spatial (three-dimensional) experience outside of school – helping dad with the car, carpentry, chemistry and meccano sets. Girls are more likely to lack experience of handling concrete objects, which leaves them at a disadvantage later for scientific work.

If girls are to get a fair chance later, it is essential that teachers not only *refrain* from further disadvantaging them, but recognise that they are disadvantaged by our sexist society, and practise positive discrimination; that is, provide girls with the mechanical and spatial experiences that boys pick up informally at play and at home.

Positive discrimination has always been practised in favour of boys. A lower eleven-plus pass mark for boys ensured that equal numbers of boys and girls went to grammar school.[20] The fact that boys perform slightly worse at tests of lingual ability has always been explained as 'late development', while girls' poorer performance at spatial ability tests was due to their 'inadequacy'.

Secondary school

Discrimination in the secondary school is more rampant. According to a DES Survey,[21] 98 per cent of secondary schools separate the sexes for some subjects. For example, girls may be required to take needlework and/or cookery (domestic subjects), while boys do handicrafts, technical drawing, etc.

Another form of open discrimination is in allocation of resources: 'Regulations issued by the late Ministry of Education prescribed standards for boys, girls, and mixed schools respectively, which gave a standard for girls' schools of fewer science labs and technical rooms than those for boys.'[22]

Areas of opportunity for girls are restricted. Just as vocational courses of further training are restricted to males, so the schools offer a wider range of optional courses to boys – such subjects as agriculture, accounts, building, catering, electronics, engineering design, surveying, fashion design. Accounts, catering, fashion design and electronics are offered to girls.

Where these courses are open to both sexes, it is more usual to find boys entering the traditionally female areas (commerce, cookery) than to find girls taking courses in traditionally masculine subjects (engineering, technology). (But then, the top jobs within any field have always been held by men – women do the cooking, but the best chefs are supposed to be men.)

In 1975, a DES survey found that 14 per cent of building and engineering courses were open to both sexes at the fourth form stage, and 45 per cent of commerce and office practice courses.[23]

Sometimes, when the school offers a supposedly 'free choice' of subjects, we find covert discrimination. For example, the DES survey showed that 27 per cent of mixed schools have pre-emptive patterns of curriculum – that is, if you don't take a certain subject in the first years, you can't go on to do a related subject in subsequent years. For example, one school offered 'technical drawing' as a free option. But only those who had previously taken metalwork could take it. Guess who had to do needlework and home economics in the first years when the boys did woodwork and metalwork?

Another factor is premature specialisation. Already in the first years, if girls don't do science they can't go on to study it later. If they take only biology O level – a common choice for girls – they can't go on to take sciences at A level. Many girls are dissuaded from studying science at an early age, by parents, peers or teachers; and by the time they reach an age to make a conscious choice, it is too late.

Even supposing the school offers pupils a free choice of subjects, girls are less likely to choose sciences. In 1975 only 15 per cent of girls took maths/science subjects at A level, as compared with 38 per cent of boys. It is ironic that, although girls are more likely to be *offered* science subjects in mixed schools, they are more likely to choose and succeed in science subjects in single-sex schools, although the standard of teaching and the facilities may be worse. Many even have to travel to nearby boys' schools which have appropriate facilities. The fact is, more girls in single-sex schools study physical sciences, whereas fewer boys in boys'-only schools study physical sciences.

This is probably because peer group pressure to conform to sexual stereotypes is greater in mixed schools. Boys do traditionally male subjects and girls do female subjects. Biology is the 'female' science, the only science that it is acceptable for girls and women to study, presumably because (1) its methods are considered easier than chemistry and physics; it involves less mathematics; (2) its content is considered suitable for girls and particularly suited to their interest; it is to do with animals and human beings. Boys are supposed to be more interested in 'things' and girls in 'people'.

Table 5 shows the percentage of pupils offered, and those taking, different science subjects in mixed and single-sex schools. The differences are small but significant. Girls in mixed schools are more likely to be offered physics, but less likely to take it, than girls in single-sex schools.

Table 5:[25] *Options in fourth and fifth forms*

	Offered		Taking	
	Single-sex	Mixed	Single-sex	Mixed
	(percentage of totals of pupils)			
Physics	%	%	%	%
Boys	85	91	51	47
Girls	62	75	14	11
Chemistry				
Boys	81	79	29	28
Girls	75	78	20	17
Biology				
Boys	79	91	31	27
Girls	96	96	4	51

Dale's[24] study of co-education shows conclusively that co-educated boys tend to show greater interest and higher attainment in maths and science than boys educated in single-sex schools. Their all-round attainment in general subjects was also higher. Co-educated girls, on the other hand, tend 'to draw away from science and mathematics because the boys are better. In general attainment the raw[26] scores are slightly but fairly consistently in favour of girls' schools.' This polarisation affects girls' chances of getting into university: 'almost twice as many girls from single-sex schools as from mixed schools, in proportion to the number of girls of each type, went on to read maths or science at university' (p.168).

Dale admits that the science findings are 'strongly in favour of the girls' schools' but is willing to dismiss them since the 'other advantages [of co-education] are so varied and so pronounced'. For the boys, he means. It is perfectly obvious that co-education benefits the boys at the expense of the girls:

> boys do better at maths than girls do, and it also tends to be regarded as a boys' subject, therefore the boys' expressed liking for the subject is *increased* by the presence of girls in the class . . . boys are spurred on by friendly competition [sic] with the girls and some of the girls' greater industriousness and conscientiousness is communicated to the boys . . . surely we should opt for co-education on the basis of rather better academic progress (boys', at least) [*sic*] . . . progress does not suffer and in the case of boys is probably improved.

It is quite obvious that Dale considers the education of boys to be more important than that of girls. What do the girls get out of co-education? According to Dale, the 'feminine sex' is bound to benefit from the male presence. Girls gain from the 'unconscious absorption of the boys' approach to mathematical problems'! All of Dale's figures belie these arrogant assumptions. His whole study shows that girls do worse at mathematics in mixed schools.

I quote at length because this study is a good example of sexist social science, where the interests of girls are considered much less important than the interests of boys.

Applying for university

Even supposing Alice manages to take science A levels at school, she's not going to find it easy to get into university. Not unless she can do *better* than the boys.

According to Phillips,[27] in 1969, pupils who took science subjects at O level were more likely to stay on at school and successfully complete a sixth-form course (50 per cent as opposed to 30 per cent of the non-science group).

School-leavers with three science A levels are more likely to get into university than those with three arts A levels. Those with mixed subjects – combinations of arts and science – have even less chance than arts specialists. These applicants are usually girls with biology and non-science subjects. Universities are very conservative: only the traditional subjects and combinations are acceptable. Boys with only two science A levels are also more likely to get into university than boys with only two arts A levels. But girls with two science A levels are less likely than either to get in.

Since science departments are unwilling to accept girls, even with the necessary qualifications, girls have to do better than boys to get the same opportunities (see Table 6).

Table 6:[28] Percentage of applicants admitted to university*

| | Three or more A levels | | Two or more A levels | |
	Men	Women	Men	Women
Arts A	60	70	53	60
Science A	79	75	67	50

*The percentage of girls may seem high, but remember they are the percentage of girls applying for university who got two or three science A levels – and there aren't very many of them.

Applying for apprenticeships

More women are applying for traditionally male fields: *Spare Rib*, in July 1976, ran an article about an electrical engineering apprentice at Portsmouth Naval Base Dockyard. In 1975, only one out of sixty applicants for apprenticeship places was a girl; in 1976, apparently since the Sex Discrimination Act, one in eleven applicants was a girl. Has the Act ensured that women are no longer discriminated against, if they are determined enough to apply?

Discrimination may be less overt, but it still operates in many ways. For instance, boys are more likely to have had relevant practical experience in handling tools and learning how machines work. Applicants for TOPS courses are expected to have some knowledge of carpentry, plumbing and motor mechanics before they start. Boys are more likely to have studied the subjects at school, practised them at home, and may even have had work experience as a plumber's mate or in a garage. Unqualified women are unlikely to have had such experience, so they are disadvantaged from the start.

They may be refused entry to the course for this reason, and even if accepted they are generally watched more closely, and are likely to fall behind because they lack a lot of informal knowledge.

Other forms of discrimination are less subtle. One woman, refused a place on a motor mechanics' course, was asked at interview if her boyfriend wouldn't object to her smelling of oil!

A woman who applied for plumbing was required to undergo a whole day's assessment. She was convinced this was because she was a woman, but they assured her that 'they'd done it before when they had an almost totally deaf man apply for a course which involved noisy machinery. . . . So it seems as though they regarded being a woman as equivalent to having a physical disability!' At the end of the day, the assessor said, 'try as I might, I can't find any reason why you shouldn't be allowed on the course. I didn't think you could do it, but it looks as though you can.'[29]

More girls are applying, but not necessarily accepted.

Applying for jobs

As I have already pointed out, although women make up 10 per cent of persons with scientific and technological qualifications, they make up less than 2 per cent of managers, practising research scientists, engineers and technologists.

The largest proportion of qualified women (36 per cent) are schoolteachers. Although the socialisation of girls might be expected to encourage girls to 'choose' this occupation, there is evidence to show that women are pushed into teaching because of limited opportunities elsewhere.

A survey of Edinburgh University[30] entrants to the science faculty showed that 21 per cent of women wanted to teach, and 57 per cent ended up doing so. Seventeen per cent of men wanted to teach and 24 per cent ended up doing so. According to the 1966 Census, 35 per cent of qualified women scientists are 'not in employment'.[31] They are, in fact, unemployed, although not necessarily registered as such. Fifty per cent of married women are unemployed. Perhaps some prefer not to go out to work, but it is likely that others are unable to because of the difficulty of combining demanding professional work with domestic responsibility and child care. Lack of nursery facilities and the scarcity of part-time work in science are partly to blame, and it is obvious that the hours and holidays of teaching make it an attractive – perhaps the only – option. School teaching used also to be one of the few occupations that you could leave, and return to once the children were at school.

Competition in science is fierce, and the advances made in only a few years may render a woman's qualifications and experience obsolete if she 'drops out' to have children. She will also have dropped behind in the production of published papers. The age-related salary scales of higher academia and the civil service make employers even more reluctant to take on an older woman.

The fact that 10 per cent of unmarried qualified women are unemployed as compared with only 0.5 per cent of qualified men seems to reflect the prejudice that 'women are not worth taking on because they're bound to get married and drop out anyway'.

Those women who find employment in science are likely to stay on the bottom rungs of the ladder. Science is a male club,[32] and women are excluded from the informal contacts and sponsorship systems of male scientific circles.

But this is not very surprising. Within any professional field there are few women, and even fewer at the top. Women are not the innovators or decision-makers in this society; most of the power positions are held by men.

Women opt out and are pushed out of science at about the same rate as out of other professional fields, and for the same sort of reasons. What is special about science is that there are spectacularly few women in the first place.

WHY GIRLS DON'T WANT TO DO SCIENCE

Girls don't want to do science partly because they are brought up to think it's a male subject, and partly because it *is* a male subject.

The image of femininity

Girls are socialised into believing that they can't do science, that it requires skills and characteristics that girls don't have: abstract thought, practical ability, self-confidence, independence, high intelligence, etc.

Girls may also be afraid of being thought 'unfeminine'. An American girl who was sent to a camp for children gifted in mathematics pretended she was going for remedial help, since it was more acceptable to her peers for a girl to be 'thick' at maths.[33]

These problems can be dealt with only if parents, teachers, career officers, employers and universities not only refrain from discouraging girls from choosing science, but actively encourage them to do so, bearing in mind that girls are at a serious disadvantage because of the image of femininity that is forced upon them.

But then we come up against the problem that science, as it is practised now in the West, is a male subject; and even if more girls realised that they could study it, they might not want to do so, for very good reasons.

The image of science

The answers to an A level question about why more students are choosing arts and social studies rather than natural sciences reveals the attitudes of sixth-formers towards the image of science.[34] Both boys and girls believed that science requires a high level of intelligence, but there was diversity of opinion as to the other important factors. Girls were more inclined than boys to blame lack of resources in science teaching, and early specialisation in arts subjects. Boredom in lessons and lack of practical work also affected more girls. Slightly more girls than boys wrote of the impersonality and inhumanity of science. Scientific knowledge, although with potentially good applications, was regarded as having been misused, and our society as being mechanistic with little concern for the individual. Forty per cent of girls (as opposed to 25 per cent of boys) expressed a desire to help the community directly and felt this would be possible through the social sciences.

These girls felt science to be too abstract, academic, impersonal and divorced from its social context. As we have already seen, attempts to make science education practical and to relate it to everyday life are generally restricted to the 'lower-ability' children, those who are not intended to specialise in academic subjects. Practical, applied knowledge has low status[35] in our society, not because it is not valued and valuable but because it is a preparation for badly paid, low-status jobs. Changing the image of science would make it more attractive to girls, but if there were two separate curricula – 'science and society' courses for the non-academic children, and O and A level, abstract, academic science for the future specialists – these methods would not produce any more women scientists. 'Science and society' courses in the early years of secondary school may encourage more girls to take up science, but the curricula in the later years would also have to change.

Universities and businesses would have to stop demanding that the training of future scientists start at school with specialisation at fourteen years of age or earlier. A broad, general education in science and its applications for all pupils would both attract more girls to the subject and make science more accessible to the non-specialist. Making scientific principles understandable to all school students would be the priority, rather than selecting and preparing the future élite.[36] The more academic students could pursue some topics in more depth, instead of following a totally separate curriculum, thereby helping to make science education less abstract and élitist and to relate it to its social context.

These changes may reduce the mystification of science as a school subject and make it more accessible and attractive to the majority of pupils. But science education only reflects the face of science itself, and it is quite definitely a male field. For a start, it is a male field simply because it is dominated by males. One of the reasons women don't want to study science or work in scientific jobs is because it is unpleasant to be in such a small minority. Think what it means to be one woman among fifty men in an engineering course at college when many of your colleagues and possibly your instructor are hostile to women, think they can't do the job as well as men, and are watching for you to make a mistake. 'The attitude of many of the staff is that they are doing women a favour letting them on the [TOPS] courses; it's then up to an isolated woman in a hostile environment to prove that she can do the job.'[37]

Another woman scientist comments:

> I have been aware of the subtleties of the prejudice against women . . . they are not taken seriously, they have to push to be heard, and if they do not or cannot push, no one listens. The prejudice resulted, in my experience, in both stupid and irrelevant jokes about women and even in questions about whether my female physiology might affect my ability. . . . It isn't so much that any particular incident gets you down, it's the cumulative effect of incident after incident which grinds you down, and drains your self-confidence and motivation.[38]

And in a scientific occupation, not only would you be heavily outnumbered, but most of the men would be in positions of authority over you. Fighting for promotion usually means competing with the few other women at the bottom.

But science is not simply a neutral field that happens to be dominated by men. In a recent issue of a magazine of alternative technology, women scientists wrote about their feelings.[39] Many of them felt alienated as women within the scientific world, for a variety of reasons. Scientific knowledge is associated with values which in our society are attributed to men. It is supposed to be objective, rational and impersonal, to the exclusion of feelings, intuitions and emotions, values commonly attributed to women. Now I am not suggesting that women are not as capable as men of 'objective' thinking; rather that total objectivity is neither possible nor desirable. In fact, intuitive and subjective thought play an important part in the scientific process but they are repressed and go unacknowledged.

'Most practising scientists realise that science is not objective although they strive to make it so.'

'The myth of science is that it is an objective study of facts and therefore devoid of emotions, and so no acknowledgement is made of their presence. And there begins my alienation from science.'

Second, the scientific community is rigidly hierarchical in structure and fiercely competitive. Anyone who has read *The Double Helix*[40] will be aware what a cut-throat business it can be. While the women expressed a desire to co-operate, sharing ideas and discoveries within the community and communicating them to non-specialists, they found male scientists motivated by the desire to beat their colleagues and acquire more money and status.

'When women enter science they enter a ready-made structure based on "male virtues" and "manly games" such as competitiveness, self-assertiveness, bombastic self-importance, self-centredness – all qualities which are considered to be socially positive ones in men and negative ones in women.'

Traits that women may not be keen on acquiring!

'Science is big business . . . with too many people chasing too little money and/or status.'

'The emphasis on individual achievement is crucial. I entered science in a naive state believing in some kind of shared excitement of inquiry and discovery of solutions to problems facing people in their attempts to live together in the physical universe. But I found the emphasis was on the performance of the individual scientist.'

'My motivation comes from feeling part of something to which I can contribute; my enthusiasm comes from sharing, and my ideas are generated by those interactions. I know that I can only feel free to express my ideas in an environment that is not intimidating and judgemental, in which I know my anxieties and doubts can be freely acknowledged, and where mutual support is the key and not individual performance. Such openness is the anathema of the scientific world I entered, where emotional dishonesty is blatant under guises of reason, objectivity and abstractions, and where the social reasons for doing science are lost among the emotional needs of

Western men to achieve, perform and acquire status in the eyes of their own sex.'

'The rules are defined by men for men and are based on brotherhood and male group bonds.'

Lastly, the women felt that science is divorced from its social context. While claiming to be 'pure' and 'neutral', scientific knowledge is pursued for its own sake, regardless of its consequences and wider implications. Science attempts to understand, in order to control, subdue and exploit nature. Its effects on the social and physical world are not considered from the beginning but are often seen as inconvenient 'side'-effects to be solved by yet more science.

'Male engineers . . . don't make much connection between the work they are doing and the good of society in general.'

'Our science is largely performed in the service of those needs of profit and social control.'

'I eventually left industry due to the complete lack of social conscience of my professional colleagues.'

'I left because I felt science was alienated from people, society and life.'

It is vital that more women study science and practise it at all levels. All women need to achieve an understanding of science and technology in order to take more control over their lives. We need to have more women involved in the creative and decision-making processes, in order that science is not used against us.

There are a hundred and one reasons why women don't study and practise science, and there are as many ways of changing this situation, by attempting to prevent early specialisation, introducing 'science and society' courses for all school students, fighting for better job conditions and so on.

But we must not lose sight of the fact that science is a male field, and not only because it is dominated by males. Attracting more women into patriarchal science is only a partial solution. We don't just want more women laboratory assistants. Nor do we want women to become 'honorary men' in a science that is based on 'male' values. Equal opportunity is simply not enough. We have something much better in mind.

Science should embody the ideas of co-operation and the sharing of knowledge, both within the scientific community and with non-scientists. It should be concerned with the benefit of the community and the balance of nature. Women could play a large part in creating a truly human science.

2

Psychological Sex Differences

Janet Sayers

INTRODUCTION

The recent revival of feminism has been accompanied by a growing interest in the subject of female psychology. Women, seeking to liberate themselves from their traditional role, have turned to psychology to discover, in order to change, the psychological determinants of that role. Those opposed to feminism have, on the other hand, looked to psychology to provide a rationale for their conservatism.

There are two research traditions within psychology which bear on the subject of female psychology: the first seeks to discover the nature of female psychology, while the second seeks to determine the nature of the psychological differences between the sexes. As the first tradition focuses specifically on female psychology, so it is to this tradition that feminists, interested in psychology, have usually turned. In particular, they have looked at the answers provided by Freud to 'the riddle of femininity',[1] and much has been written about the services and disservices done to women by his account of their psychology.[2] Feminists have paid less attention to research on sex differences. Nevertheless, this research is of importance to feminism; both because it has frequently been used to shore up sexist ideology (that is, the ideology that holds that women and men should pursue different, and by implication unequal, social roles); and because it studies a subject of central concern to feminists – namely, the character and determinants of existing sex differences.

SEX DIFFERENCES AND PSYCHOLOGICAL TESTS

Understanding the general character of female (or male) psychology has certainly not been the primary objective of most sex differences research. Had it been so, researchers might have looked at the similarities, as well as at the differences, between the sexes. One reason why psychologists have focused on sex differences, rather than on sex similarities, has been that they have believed that these differences might be of use in constructing tests of psychological adjustment. The assumption underlying such a use of sex differences data has been that adjusted people conform with the behaviour that differentiates their sex from the opposite sex. These data thus appear to constitute a convenient yardstick against which to assess a person's normality or abnormality.

One researcher who used sex differences data in this way was Daniel Brown,[3] a psychologist working for the psychiatric service of the American Air Force. He devised a test, for use with children, which consisted of thirty-six pictures of things 'commonly associated with masculine or feminine roles'. These included pictures of toys (e.g., tractor, train engine, gun, soldiers, purse, doll, cradle, dishes), and pictures of objects presented in pairs (e.g., sewing materials–aeroplane parts, mechanical tools–household objects, building tools–baking articles). Children doing the test are asked to indicate which of these toys, and which one of each pair, a character, referred to as 'It', would like. Test scores are obtained by giving one point for each 'male toy picture' and zero for each 'female toy picture' chosen.[4] Despite the lack of value given to feminine choices both in this test and in the wider society, Brown implies that, at least by adolescence, girls should choose the objects conventionally assigned to them if they are to develop normally, and avoid becoming lesbians! He says, for instance, of masculine role preference (which he equates with preference for things 'typical of and associated with the masculine role'), that if a girl continues 'to prefer the masculine role through childhood and into adolescence (and conversely with the boy) sex-role inversion in adulthood would be expected, one aspect of which would be a homosexual object choice'.[5]

While Brown simply relied on conventional wisdom about sex differences in object preferences to construct his test of sex-role conformity, other psychologists have measured this aspect of behaviour by gathering data on how the sexes actually do differ in their behaviour. Terman and Miles,[6] for instance, working for an American research committee investigating 'problems of sex', looked at the ways men and women differ in their responses to questionnaire items. They then used these data to develop a masculinity–femininity (M–F) test for use in assessing 'deviation from the mean of either sex', for researching into issues like homosexuality, and for 'educational, vocational, and avocational guidance'.

Although the research of Terman and Miles, and of Brown, was not designed to investigate the specific nature of female psychology, they did chance on some interesting facts about that psychology. Brown, for instance, found that girls, at least until they are ten years old, show a weaker preference for the feminine role than do boys for the masculine role. In other respects their work is not only uninformative, but is quite misleading about female (and male) psychology. One example is Brown's assumption that children who conform primarily with the norm for the opposite sex will become homosexual in adulthood. He provides no evidence for this assumption. What evidence there is suggests that the equation of homosexuality with such cross-sex behaviour is, in fact, often false.[7]

Brown, and Terman and Miles, also implicitly assume that the more feminine one is in one's behaviour the less masculine one must therefore be (and vice-versa). This assumption is built into their tests in that each feminine choice on their tests automatically reduces one's possible masculinity score. The possibility of registering both feminine and masculine choices in response to any one item is also excluded – the child doing the 'It' scale cannot, for instance, express a liking for both baking articles and building tools. The tests are premised on the belief that femininity and masculinity can be represented as lying at opposite ends of the same continuum, that femininity is the reverse of masculinity.

Recent work by Sandra Bem[8] at Stanford University shows this belief to be false. She asked students to indicate how well each of a number of personality characteristics applied to themselves. The list of characteristics included traits regarded by Bem and her students as feminine (e.g., compassionate, loves children, gullible), and traits regarded by her and her students as masculine (e.g., aggressive, independent, ambitious). Bem found that the extent to which an individual endorsed feminine items could not be predicted from the extent to which she endorsed masculine items, that femininity in an individual does not preclude that individual also being masculine, and that femininity and masculinity cannot be validly represented as lying at opposite ends of a single dimension. She also discovered that a sizeable minority of subjects not only combined masculine and feminine characteristics but combined these characteristics in approximately equal degree – that 27 per cent of the women and 34 per cent of the men in her sample could be classified as 'androgynous' in this sense. By providing separate measures for femininity and masculinity, Bem thus rediscovered what had already been known for some decades to psychoanalysis, namely that people combine, in varying degrees, both masculine and feminine aspects in their behaviour.

Freud[9] arrived at this discovery by investigating, rather than assuming, the conventional equation of female psychology with femininity, of male psychology with masculinity. His clinical work convinced him that this equation was, in fact, invalid. Girl and boy infants, he said, manifest masculine and feminine traits in approximately equal degree. Furthermore he maintained that, although the sexes do subsequently come to differ in the balance of these traits, they nevertheless retain the legacy of infantile 'bisexuality' even in their adult behaviour.

Freud was also of the opinion that it was unhealthy for women to suppress all their masculinity in the process of developing their femininity.[10] More recently Bem has arrived at a similar conclusion. She found that 'feminine' women (i.e., women who, on her test, endorsed feminine items and rejected masculine items as not applying to themselves) were less well adjusted – at least in terms of their ability to deal effectively both with situations demanding 'masculine' independence and with situations demanding 'feminine' nurturance – than were 'androgynous' women (i.e., women who, on her test, endorsed both feminine and masculine items in approximately equal proportions).[11] This finding casts doubt on the validity of another assumption of the 'It' and M–F tests, namely the assumption that singleminded conformity with the traits typical of one's sex guarantees psychological adjustment. If one believes that the sexes should pursue different social roles, and that femininity fits women for one social role while masculinity fits men for a quite different social role, then it follows that women should be feminine and men masculine if they are to adjust to their roles in society. The equation of femininity in women, and of masculinity in men, with psychological adjustment may serve sexist ideology, but it is not valid in terms of the actual facts regarding female and male psychology.

Apart from shoring up certain false assumptions of sexist ideology, sex differences research for the purposes of constructing M–F tests has also trivialised the subjects of male and female psychology.[12]

Terman and Miles, for instance, chose items for their test on the sole basis that women and men tended to respond differently to them. This certainly made their task of selecting test items relatively easy. Since, however, items were not selected on the basis of existing knowledge and theory about the important aspects of female or male psychology, there was no guarantee that they would have anything but a trivial bearing on it. And, indeed, many of the items in their test are quite absurd on any account of female and male psychology. One could include here, for instance, the following items:

'Are you much embarrassed when you make a grammatical mistake?'[13]
'Can you usually sit still without fidgeting?'
'Do you hear easily when spoken to?'

The trivialisation of psychology in the interests of developing tests for use in clinical, educational and vocational guidance is not confined to M–F tests. In developing intelligence tests, for instance, psychologists have chosen test items on the basis that they differentiate, in an appropriate statistical fashion, between older and younger children, or between more and less able children and adults. This certainly facilitated their job of selecting test items. The resulting intelligence tests also clearly served, and continue to serve, the function for which they were devised: namely, that of channelling children into different kinds of schools,[14] and of channelling adults into particular jobs. Since, however, the items in these tests were not selected with reference to existing knowledge and theory about intellectual function, or about the development of intelligence, they often bear little or no relation to the important aspects of intelligence.[15] Intelligence tests have, accordingly, often earned psychology the disrespect of many of those members of the public who have been exposed to them.

SEX-ROLE STEREOTYPES

Developmental psychologists (that is, psychologists who study the ways individuals change psychologically as they grow older) have had somewhat different reasons from the psychological testers for studying psychological sex differences. Like feminists, they have been interested in discovering how these differences come to develop in children. But whereas feminists have been interested in this subject because they have wanted to discover how these differences might be changed or reduced, psychologists have often used their 'expertise' on this subject to advocate that these differences should, in general, be maintained. Biologically-minded psychologists have regarded conformity with existing sex differences as part of our duty to evolution.[16] On the other hand, psychologists who explain these differences as the product of socialisation in childhood have, like the psychological testers, advocated conformity with these differences in the interests of mental health. Jerome Kagan, who has been involved in one of the largest projects to investigate the development of sex differences, suggests that the goal of normal development should be that of acquiring 'the culturally approved characteristics' for one's sex. He claims (without providing any supporting evidence) that those who 'strive to avoid adoption of sex-typed responses . . . are typically in conflict and are likely to manifest a variety of psychopathological symptoms'.[17]

The study of psychological sex differences has thus been used by developmental psychologists, as it has been used by the psychological testers, to urge conformity with existing sex differences. Developmental psychologists, unlike the psychological testers, have, however, focused much more consistently on characteristics generally thought to be important in male and female psychology. They have looked for sex differences in, for instance, aggression, dependency, achievement-motivation, suggestibility, and sociability. That these are the kinds of characteristics that are commonly believed to differentiate the sexes psychologically is indicated by research into the content of sex-role stereotypes. One such study[18] found that over 60 per cent of five-year-old children expected men to be more 'aggressive', 'independent' and 'ambitious' than women, and expected women to be more 'emotional' and 'soft-hearted' than men. Another study[19] showed that over 70 per cent of college students expected men to be more 'competitive', 'logical' and 'worldly' than women, and expected women to be more 'easily influenced', 'passive', and 'aware of feelings of others' than men. Instead of questioning the validity of these expectations, however, psychologists have tended to assume their validity and then interpreted their data accordingly. They have often simply sought data to corroborate sex-role stereotypes, rather than systematically examining whether the sexes actually do differ in the ways assumed by the stereotypes.

An experiment by Susan Goldberg and Michael Lewis,[20] an erstwhile co-worker with Kagan, is typical in this respect. They looked for sex differences in dependency behaviour among thirteen-month-old children. They chose to measure this trait by seeing how close the child stayed to, and how much it sought help from, its mother in an experimental playroom. On both measures girls appeared to be more dependent than boys. Goldberg and Lewis therefore concluded that girls are already more dependent than boys in infancy; that, even at this early age, girls already show what these psychologists refer to as 'sex-role appropriate' behaviour.

Had Goldberg and Lewis been more concerned to investigate, rather than to bolster, prevailing opinion on this matter they might have measured dependency more thoroughly. Psychologists who have bothered to compare the results of using different measures of this trait find that, in fact, there is no consistent tendency for girls of this age to be more dependent than boys.[21] Those who have measured dependency, for instance, by the extent to which the child protests when its mother leaves the room have found either no sex difference, or a difference indicating greater dependency in boys. Unfortunately, however, these studies remain either unpublished,[22] or buried in the recondite journals of academic psychology. It is only the findings that corroborate sex-role stereotypes that have been generally reported in the more popular and widely available books on psychology.[23]

A similar situation holds regarding research into sex differences in sociability. Stereotypically girls are thought to be more interested in social relationships and more responsive to them. The research findings that accord with this stereotype have often been reported in the more accessible books on psychology. Corinne Hutt,[24] for instance, cites such research in her much-read paperback on psychological sex differences. She says that the research shows that boys are interested in things, girls are interested in people. Until recently the major source of evidence for this kind of conclusion came, not from examining the degree to which boys and girls actually interact socially with other people, but from looking at the kinds of books they liked to read, the sorts of films they liked to see, what they talked about, and the kinds of subjects they depicted in their drawings. On the basis of this indirect evidence one very influential, academic review of the sex differences literature concluded:

> In psychological development, *from earliest infancy*, males exhibit a greater interest in objects and in their manipulation, whereas females show a greater interest in people and a greater capacity for the establishment of interpersonal relations.[25] [my emphasis]

Hutt provides some direct evidence for this conclusion when she reports finding that nursery school girls spend more time in social interaction, while nursery school boys spend more time in physical activity. However, the vast majority of studies of peer group interaction among nursery schoolchildren indicates either no sex difference, or that boys are more sociable, in this sense, than girls at this age. Studies of school-age children have found that, at this age too, boys make friends with more children, and are more influenced by other children, than are girls. [26] These findings may be consistent with the stereotype of boys as being more 'worldly' than girls, but it hardly corroborates the stereotype of girls as being more 'sociable' than boys. Far from examining the contradictions between these different aspects of sex-role stereotypes – in this case, between women's assumed sociability and their lesser degree of worldliness – the popular reporting of research on sex differences has been more concerned to substantiate each aspect of the stereotypes in turn without looking at how these several aspects relate to each other.

The willingness of psychologists writing general books on psychology to report only those few studies that corroborate sex-role stereotypes, and to imply, on this flimsy basis, that the sexes do differ in the manifold ways in which they are stereotypically expected to differ, is often quite blatant. John Nash, for instance, introduces his chapter on psychological sex differences by admitting that 'some of the studies to be mentioned in the following review are unreplicated, and even where there are related studies, strict comparison of results is commonly lacking.' [27] Although scientists are usually cautious about accepting the results of any one experiment before it has been repeated in other comparable situations, John Nash shows no such caution about the 'unreplicated' data on sex differences. Instead, he emphasises, on the basis of these data, that 'one thing that does become apparent is that males and females differ in a surprising variety of ways'. [28]

As we have seen, however, where writers have bothered to compare 'related studies', they have found that the sexes do not differ in quite all the ways predicted by sex-role stereotypes. Adults apparently agree in expecting the sexes to differ in over forty different ways! [29] A recent very comprehensive review [30] of the data on sex differences concludes that psychologists have, in fact, convincingly demonstrated only four psychological differences between the sexes. Eleanor Maccoby and Carol Jacklin, working, like Sandra Bem, at Stanford University, claim that the only differences that have been clearly demonstrated are that boys are more aggressive, and are more skilled in visual–spatial and in mathematical abilities than girls, and that girls are more capable verbally than boys. Still more recent evidence indicates that even these few differences might not be conclusively established.[31]

SEX DIFFERENCES AND SOCIAL ROLES

In summarising their review of available sex differences research, Maccoby and Jacklin conclude that 'there are many popular beliefs about the psychological characteristics of the two sexes that have proved to have little or no basis in fact'.[32] Given that these beliefs remain largely unproven, why have they remained so pervasive? The answer lies, at least in part, in the ideological function served by such beliefs. Beliefs about psychological differences between the sexes are frequently used to justify the different places occupied by men and women in society. The anthropologist, Marilyn Strathern, maintains that sex-role stereotypes serve this function not only in our own society, but also in non-industrial societies. She says: 'The achievement of gender stereotypes . . . is to make it appear perfectly "natural" that men should be better suited to some and women to other roles.'[33] In our own society social commentators have repeatedly adduced psychological differences between the sexes in seeking to prove that men and women should pursue different social roles. In the nineteenth century, for instance, the prominent political theorist, Walter Bagehot, cited such differences in the hope of demonstrating the folly of those who, like John Stuart Mill,[34] pleaded for an end to the 'subjection' of women. Bagehot told his readers that throughout the animal kingdom the sexes differ in boldness, pugnacity, adventurousness, restlessness, mildness, gentleness and inoffensiveness; that these characteristics fitted males and females for the different tasks assigned them by nature; and that

> the attempt to alter the present relations of the sexes is not a rebellion against some arbitrary law instituted by a despot or a majority – not an attempt to break the yoke of a mere convention; it is a struggle against Nature . . .[35]

Similar arguments are advanced today for continuing the traditional division of labour between the sexes. Nowadays, however, professional psychologists have added their weight to these arguments, claiming that their scientific expertise on the matter of sex differences gives them the right and duty to comment on sex roles and changes in them. John Nash, for instance, says: 'One function of the psychologist as scientist is to record such changes, but he also has the obligation to comment on them and judge them.'[36] In fulfilling this 'obligation' Corinne Hutt uses data on psychological sex differences in the same way as the Social Darwinists of the nineteenth century:[37] namely, to justify the traditional view of the proper division of labour between the sexes, and to criticise attempts to change it. She argues that males are 'more aggressive, more exploratory, more vigorous and group-oriented' than females, and that females are 'more nurturant, more verbal, and more concerned with morals and social conventions' than males. She claims that these sex differences were initially evolved to fit men and women for their different social roles in hunting and gathering societies, that they continue to fit the sexes for different social roles today, and that 'A woman's primary role is that of motherhood and most women have some or other of the attributes which fit them for this role.'[38] She concludes that attempts to change the social roles of the sexes are misguided, and says that 'It would be a pity indeed if women sought to make this less of a man's world by repudiating their femininity and by striving for masculine goals.'[39]

The assertion that sex-typed attributes fit women and men for different social roles may serve an ideological, and conservative, function in legitimating existing differences in the social position of the two sexes. Since, however, it is made in the virtual absence of any research into the actual traits needed to fulfil particular sex-typed social roles, it cannot lay claim to be anything more than a statement of belief. The assumption that femininity fits women for one social role and masculinity fits men for a different social role is certainly not based on scientifically established fact.

One exception to the general failure to investigate empirically the bearing of sex-typed traits on the performance of particular sex-typed social roles is a study by Nilsson.[40] He found that 'feminine' women (i.e., those who assessed themselves as gentle, emotional, dependent, etc.) reported more psychiatric symptoms during pregnancy than did 'masculine' women (i.e., those who assessed themselves as strong, self-confident, etc.). This finding suggests that, contrary to sexist ideology, 'feminine' sex-typed traits do not necessarily fit women well for their social role, at least in so far as this role includes maternity. Nilsson, however, seems to have been more committed to the validity of sexist ideology than to drawing the obvious conclusions from his data. He argued that the 'masculine' women wished to appear 'healthy' and therefore denied their symptoms. He thus implied that they were less well adapted to maternity than the 'feminine' women, that these data, far from throwing sexist ideology into doubt, vindicated its belief in the fit between psychological femininity and the female role in motherhood.

Although Nilsson can be accused of distorting the interpretation of his data in the service of sexist ideology, his findings do at least constitute a test of that ideology in so far as they bear on its assumption of a fit between femininity and maternity. Similar investigations have not been conducted to determine whether the supposedly masculine trait of aggression fits men for their social role, although Corinne Hutt assures us that this trait, among others, is 'adaptive in terms of the reproductive roles that males and females fulfil'.[41] She asserts that males are biologically more prone to aggression than females and cites the following findings in support of her assertion.

1 Male rats show more 'biting', 'threat' and 'bristling' behaviour than female rats when subjected to isolation or electric shock.
2 Male infant monkeys show more 'threat', 'rough-and-tumble', 'wrestling, rolling and sham-biting' responses than females.
3 Preschool boys are 'twice as aggressive' as preschool girls.
4 In 'all societies the delinquent and criminal populations are predominantly male'.[42]
5 Aggressive behaviour is, at least in part, controlled by 'male hormones'.

It is not at all clear, however, how the greater aggression of preschool boys, or the propensity of males for delinquent and criminal behaviours – let alone the greater aggression of male rats and monkeys – bears on the performance, by men, of their 'reproductive role' in this society. Hutt relies on ideology, not on any relevant factual evidence, to supply the link here between the above data and the male role in contemporary human society.

Sex differences in intellectual traits have also been said, by some psychologists, to fit men and women for their different roles in reproduction. Verbal abilities generally appear to develop earlier in girls than in boys. In adolescence, too, girls seem to do better than boys on both 'high-level' verbal tasks (e.g., on tasks requiring reasoning by analogy, comprehension of difficult written material, and creative writing) and on 'lower-level' measures (i.e., of verbal fluency).[43] Two Oxford psychologists, Jeffrey Gray and Anthony Buffery,[44] argue that this sex difference is the product of evolution. They suggest that verbal skills were evolved to a greater degree in women than in men because they fitted women for their role in child-rearing, for their task of teaching children to speak. These psychologists thus imply that women are generally better fitted than men to the task of child-rearing; that the current situation in which women overwhelmingly do perform this task is well justified; and that attempts to share this task more equally between the sexes are misguided. The verbal superiority of women could, of course, equally be cited to question the wisdom of the situation in which men vastly outnumber women in occupations requiring verbal skills. Such, however, is Gray and Buffery's commitment to using their data to justify, rather than to question, the existing division of labour between the sexes that they do not even consider this possibility.

Psychological sex differences have also been cited to justify sex differences in occupational life. Corinne Hutt, for instance, claims that the greater manual dexterity of women fits them for being 'seamstresses and needlewomen', and for being 'extremely competent typists'; and that the 'feminine facility . . . for rote memory' fits women for secretarial work.[45] She thus implies that the existing preponderance of women in all these occupations, at least in the West, is fully justified.[46] Psychological sex differences that do not accord with existing sexual inequalities in the work setting tend to be ignored. Helen Thompson's[47] finding that in a finger-tapping task boys tap faster than girls has never, to my knowledge, been cited to question whether the present preponderance of women as typists in our society is justified.

The position, or rather the lack of position, of women in the scientific laboratory has also been attributed to their supposed inferiority to men in the traits necessary for scientific work. James Swinburne told his readers in 1902:

> When we come to science we find women are simply nowhere. The feminine mind is quite unscientific. Men are curious about things, women about people. . . . Many women can do some sort of scientific work. They are more careful than men, and more accurate in taking readings. In this direction they make excellent assistants, and they could probably do the routine work of the assay, or city analyst's office, or of an observatory, better than most men. Ladies' names often appear as authors of papers, generally in organic chemistry, or in subjects involving tedious but accurate readings of instruments, but such work is either done in conjunction with men, or is obviously under their guidance and supervision, and much is made about it out of gallantry.[48]

More recently, Maccoby and Jacklin have suggested that one cause of women's lack of scientific achievement may be their inferiority to men in spatial ability; women apparently do better than men on physics tasks requiring verbal skills, but worse than men on physics tasks requiring visual–spatial skills. Maccoby and Jacklin suggest that sex differences in spatial ability might well be a product of cultural factors. They point out that no such difference is found among the Eskimo, where boys and girls are both granted 'considerable independence', whereas substantial sex differences are found on this trait when 'males exercise strong authoritarian control over females'.[49] Other psychologists have cited evidence indicating that sex differences in spatial skills might be due to differential rates of brain development in boys and girls. Some of these latter psychologists[50] have then used these data to suggest that women's under-representation in certain professions, including certain scientific professions, is biologically determined. They thereby neglect, and divert attention away from, external factors – the systematic discouragement of girls from pursuing scientific interests and careers, sex discrimination in job recruitment, and practical difficulties concerning child care, etc. – which also play a large part in obstructing women's entry into the scientific professions.

Just as the data on sex differences have been used to justify existing inequalities in the social position of women and men, so they have been used (as indicated above, pp. 50–52) to combat attempts to reduce these inequalities. Psychologists have also cited sex differences in questioning the advisability of giving women equal opportunities with men in professional life. Max Coltheart,[51] a psychology professor at London University, argues that, since boys perform better than girls on visuo-spatial tasks (like the one illustrated in Figure 1), they will 'obviously' tend to make better architects or engineers. This is to ignore the many factors, other than the ability to do these tests, that are also involved in becoming a good architect or engineer. Nevertheless, Coltheart concludes, on the basis of such limited test data, that any policy that seeks more nearly to equalise the recruitment of women and men to the architectural and engineering professions would be ill-advised.

If Coltheart, Hutt and others have questioned the principle of equal opportunities for women and men in adulthood, even more psychologists have used sex differences research to question the advisability of equal education, even of co-education, for girls and boys. John Nash concludes a chapter on sex differences in his textbook on psychology by stating that:

> these differences require them [children] to have distinct child rearings or, at least, to be reared in manners more distinct than is customary in our society . . . we should begin to look seriously at our co-educational system of schooling and our indistinctly differentiated child rearing practices and ask if they really meet the developmental needs of girls and boys.[52]

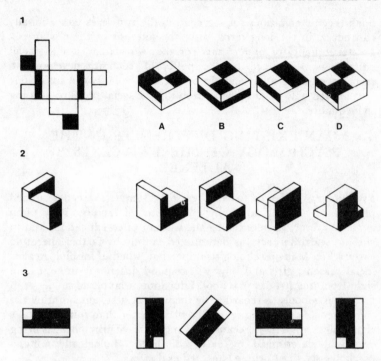

Figure One: *An example of psychological tests of 'spatial ability'*

In panel 1, choose which of the four figures labelled A, B, C, and D could be made by folding up the object on the left-hand side of the panel. More than one answer may be correct. In panel 2, choose which of the four objects on the right is the same as the object on the left, just viewed from a different angle. In panel 3, choose which of the four figures on the right could be produced simply by rotating the left-hand figure. (answer on next page)

The recommendations of psychologists regarding co-education are not simply confined to the textbooks. They are frequently reported in the press. Thus, for instance, a recent newspaper article tells us that:

> if you want your daughter to be a physicist or an engineer, your best plan is to send her to an all-girls' school. The experts suggest two reasons. First, since the girls' spatial and mechanical skills develop less quickly than the boys' they become discouraged when they work alongside their male peers. Second, the sex-role pressures are stronger in mixed than in single-sex schools and a girl is, therefore, more likely to be afraid of opting for science and thereby appearing 'masculine'.[53]

Such recommendations for segregating the two sexes educationally can never, in the long term, serve the interests of those who seek greater equality of opportunity for the two sexes,[54] just as racial equality is not served by the policy of 'separate development' recommended by apartheid. In both cases, equality can ultimately be achieved only by directly confronting the social pressures leading to inequality.

CAN EXISTING DIFFERENCES IN THE PSYCHOLOGY OF THE TWO SEXES BE ALTERED?

Psychologists differ as to whether sex differences are determined primarily by biological or by environmental factors. They often agree, however, in questioning the wisdom of co-education, and in viewing sex differences as determined mainly by factors operating early in life. Many psychologists imply that, whether for biological or social reasons, girls and boys will respond differently to education right from the first day of school. Education, whether in single- or in mixed-sex schools, is regarded by them as, at best, ameliorating the consequences of existing sex differences, rather than fundamentally changing them. This is because they regard these differences as being essentially determined by early childhood. Biologically minded psychologists, like Corinne Hutt, tell us that:

> The evidence strongly suggests that at the outset [of life] males and females are 'wired-up' differently. Social factors thus operate on already well-differentiated organisms – predisposed towards masculinity or feminity.[55]

As we saw earlier (p. 45 above) Freud located the origins of psychological sex differences somewhat later in development, namely in late infancy. He suggested that the character of this differentiation is largely determined by the psychological effects on girls and boys of the discovery of the genital differences between them.[56] Environmentally minded psychologists, on the other hand, suggest that social, rather than biological or genital, factors operating in infancy are crucial in determining the psychological differences between the sexes. Money and the Hampsons[57] have even suggested that what happens in the first eighteen months of life determines whether the child thereafter adopts masculine or feminine characteristics in its behaviour. They base this conclusion on their work with patients

Answers to Figure One: D; D; B and D.

whose biological sex is ambiguous – that is, patients for whom the biological indicators of sex (i.e., external genitalia, internal repro-ductive structures, gonads – ovaries or testicles, hormones and chromosomes[58]) are inconsistent, and do not clearly indicate one sex or the other. The psychological sex (i.e., gender) of these patients appears to correlate less with their biology (in so far as this indicates one sex more than the other) than with the sex to which they were assigned at birth. Money and the Hampsons suppose that it is the ways in which children are treated differently according to their assigned sex, rather than biological factors, that determine their gender identity. Usually, they say, by the age of eighteen months such socialisation processes have essentially determined the child's gender, since attempts to change it subsequently (so that it more nearly conforms with the child's biological sex) are usually unsuccessful.

These various writers imply that the psychological characteristics generally regarded as typifying female (and male) psychology are established by early childhood and cannot be substantially changed thereafter. This suggestion is clearly convenient to those who wish to maintain existing social role distinctions between the sexes, and who justify these distinctions in terms of the claim that they are based on virtually unmodifiable differences in the psychology of women and men.[59] If one were seriously concerned to investigate the validity of such a claim one could examine whether the so-called 'typical' features of women's, as contrasted with men's, psychology can, or cannot, be changed after infancy, perhaps as an effect of alterations in the social environment. However, this kind of investigation has, until recently, barely been pursued by psychologists, although they had almost unanimously implied that such psychological changes could not be effected in the behaviour of women (or men). Despite the pronouncements of psychologists on this matter, feminists have been optimistic that even grown women could come to change psychologically through political action, and through participation in consciousness-raising groups. Given this lead by developments within feminism, some psychologists have begun to investigate whether participation in such groups, or in women's studies courses, does change the psychology of the women involved in them. They find that such experiences do indeed have this effect – that they lead to changes in the sex-role attitudes of the participants.

A social change that has affected many more women than con-sciousness-raising or women's studies groups, however, has been the enormous expansion in the female labour force. Although the effects of this development on female psychology have been a subject of much public discussion, they have barely been investigated by psychologists. Instead, many psychologists have continued to imply that women are psychologically unsuited to occupational life, and that they are better suited to work inside rather than outside the home. Certainly there is some evidence to suggest that women are wary of occupational success, particularly of success in occupations (and they include most of the better-paid jobs) in which men are the predominant sex.[60] However, there is also evidence to indicate that women are not uniform in this respect – some women do show similar career choices to those of their male peers, and this seems to be associated with their having had working mothers.[61] One might therefore predict that increases in the numbers of working women will lead to changes in female psychology, at least in the attitudes of the next generation of women towards participation in occupational life. In the meantime psychology has yet to investigate fully whether, and how, participation in the work force changes the psychology of working women themselves.

SEX DIFFERENCES: RETROSPECT AND PROSPECT

In 1910, Helen Thompson Woolley said of research into sex differences:

> There is perhaps no field aspiring to be scientific where flagrant personal bias, logic martyred in the cause of supporting a prejudice, unfounded assertions, and even sentimental rot and drivel, have run riot to such an extent as here.[62]

The situation has barely improved today! The psychological study of sex differences continues to be guided more by prevailing beliefs about how the sexes do, and should, behave than by a concern to investigate how they *actually* behave, or to determine whether existing sex differences could be altered or reduced. This is not an .inevitable feature of such research. Freud, for instance, in his investigation of the psychological differences between the sexes, clearly distinguished between prevailing beliefs about these differences, and the actual behaviour of the two sexes. He recognised that stereotypically the sexes are expected to differ in terms of their activity and passivity. He says, for instance: 'when you say "masculine", you usually mean "active", and when you say "feminine", you usually mean "passive"'.[63] Instead of assuming the validity of this conventional equation and interpreting his observations accordingly, Freud used these data to examine the equation's adequacy and found it wanting. He concluded that:

> Even in the sphere of human sexual life you soon see how inadequate it is to make masculine behaviour coincide with activity and feminine with passivity. A mother is active in every sense towards her child; the act of lactation itself may equally be described as the mother suckling the baby or as her being sucked by it. [64]

Freud's method of work here contrasts strongly with that of Nilsson (see p. 51 above), who assumes that feminine traits (i.e., the traits conventionally expected of women) do accord with the demands of maternity, and interprets his data accordingly. By simply reiterating the prevailing belief that femininity fits women for their social role, Nilsson tells us nothing (at least in the conclusions he draws from his data) that we could not have discovered from prejudice alone. Freud, on the other hand, by using his observations to examine the validity of sex-role stereotypes, thereby advances our knowledge of female psychology beyond what the stereotypes could have told us. Not only did Freud draw attention to the fact that adult women are active, as well as passive; but he also used his clinical evidence, and data on infant behaviour, to show that such 'bisexuality' in adult women originates in infancy. The evidence shows, he said, that there are no sex differences in the balance of activity and passivity in infancy, that infants are, in this sense, 'bisexual'. Developments in later infancy, according to Freud, do lead to a preponderance of passivity over activity in female behaviour, and to a preponderance of activity over passivity in male behaviour. Nevertheless, despite these later developments, he maintained that the legacy of early infantile bisexuality is evident in the behaviour of both men and women. Sex-role stereotypes, at least as charted by academic psychology, could not have told us of such bisexuality in the behaviour of the sexes.

Unlike many psychologists who have used research into psychological sex differences for almost entirely ideological purposes, Freud also used his later discoveries about female psychology as a starting-point from which to re-examine, and re-formulate, his general theory about psychological development in infancy. [65] The academic psychology tradition of sex differences research has thus far been so dominated by prior ideological concerns that it has barely used its data to further our knowledge, as opposed to our prejudices, about female psychology, let alone to further psychological theory more generally. Nevertheless there are signs that such developments might well take place within this branch of psychology.

Many, both within and outside the women's movement, suppose that psychological sex differences must result from differences in the ways that girls and boys are brought up. Some psychologists,[66] who favour environmentalist accounts of these differences, have tended to simply reiterate this commonsense, 'differential socialisation' hypothesis without investigating how such socialisation actually takes place. Others have attempted to provide a more detailed account of the mechanics of sex-role socialisation by assimilating it to one or other of the existing general psychological theories about child development. Lawrence Kohlberg,[67] for instance, has suggested that the development of psychological sex differences can be explained in terms of what he refers to as a 'cognitive–developmental analysis'. This analysis takes as its starting-point the theoretical formulations about child psychology propounded by the eminent Swiss psychologist, Jean Piaget. According to Piaget, the child's behaviour develops as a function of changes in the content and structure of its, the child's, mental representations of its world. Kohlberg suggests that psychological sex differences similarly develop as a function of changes in the content and structure of the child's ideas about sex roles, and that these changes parallel the changes that are taking place in the child's ideas about other aspects of its world. Other psychologists, on the other hand, reject such a cognitive–developmental analysis of the development of sex differences and suggest, instead, that this development can be best explained within the general rubric of 'social learning theory'.[68] According to this theory social behaviour, including sex-role behaviour, develops as a result of observing and modelling the behaviour of others, and as a result of being rewarded (either directly or vicariously) for such imitation. In fact, however, neither of these two theories – cognitive–developmental, or social learning theory – adequately explains the data on sex differences.[69] It is therefore possible that these data may force changes to take place in these general psychological theories of development, just as Freud's later discoveries about female psychology led him to reconsider his general account of early infant development.

Many psychologists have regarded the study of psychological sex differences as quite peripheral to the main concerns of psychology. Were it to be governed less by ideology, and more by a genuine concern to understand the nature of psychological sex differences and their development, this area of research could lead to important developments within mainstream psychological theory. Not only could such research thus serve psychology, it could also serve the interests of women by providing them with knowledge about female psychology, and its determinants. Such knowledge is clearly relevant to those concerned to change the consciousness and the position of women. In so far as psychology has, implicitly or explicitly, opposed such changes, it has, of course, been inimical to feminism. Nevertheless, it has served, and will serve, women where it takes their actual behaviour, rather than prejudices and stereotypes about that behaviour, as the starting-point for its inquiry.

3

Sociobiology:
So What?

Deirdre Janson-Smith

The publication of Edward Wilson's massive tome, *Sociobiology, The New Synthesis*,[1] in 1975 released an equally massive scientific–political storm in America, and – in a paler form – in Britain. Wilson's attempt to formulate a general evolutionary theory of animal social behaviour which would encompass both ants and humans drew down on his head accusations of fascism, racism, sexism and even a jug of water. Confessedly surprised at the public reaction to his book, Wilson was led to 'read more widely on human behaviour' and eventually to produce a far slimmer volume in 1978, entitled *On Human Nature*.[2] Although my original brief for this chapter was to contribute to a feminist analysis of biological theories of human social behaviour, as typified by sociobiology,[3] and the 'vulgar' sociobiologies of Tiger, Fox, Ardrey, Storr, and Morris, I have decided instead to concentrate on *On Human Nature*, and on the portrayal of women that it contains. I do this for several reasons.

The simplest reason is one of space. Wilson's 1978 book alone is eminently quotable and provocative throughout, and raises more issues per printed page than I have space to deal with here. More importantly, *On Human Nature* will reach a far wider audience than *Sociobiology*. Advance publicity stated that the book was to be launched with $30,000 worth of promotion, and 30,000 hardback copies on first American printing alone, exclusive of book club orders.[4] Clearly, in the light of this the book has to be examined not merely on its scientific content but for its more general merits and likely impact.[5] What it does *not* say, what impression it leaves, will be as important as what it says.

As Wilson writes in the preface, '*On Human Nature* is not a work of science, it is a work about science.' Our book is concerned with the ways in which scientific knowledge is produced and used in our society, and Wilson's book offers a particular vision and philosophy of science with which we disagree. Therefore, although I will be devoting a large proportion of the chapter to a close analysis of *On Human Nature*, I relate my discussion to the wider themes of this book.

This chapter will first briefly examine the general science of sociobiology and some of the biological ideas on which it is based. The discussion will then focus on the description of the evolution of human societies, and in particular will look at the postulated 'biological basis' of the sexual division of labour. I have chosen to look at this debate both because I see it as demonstrating the general weaknesses of human sociobiology, and because it is obvious, in our society, that the question of women's 'suitability' for certain work, and the consideration of their 'natural' sphere of work, is politically important.

WHAT IS SOCIOBIOLOGY?

The glossary definition of sociobiology which Wilson gave in 1975 was: 'the systematic study of the biological basis of all social behaviour' – interestingly expanded in 1978 to read 'the systematic study of the biological basis of the behaviour of all organisms, including man'.[6] Humankind, in sociobiological theory, is to be treated as just another social species, admittedly with certain unique characteristics, but having a biologically based pattern of social behaviour. Sociobiology is a discipline within biology that uses the predictions and tenets of Darwinian natural selection theory to explain the structure of animal societies in terms of their evolutionary function.

I am going to assume that the theory of natural selection is to *some* extent familiar to readers, at least in its crudest form of 'survival of the fittest' – that an animal's first imperative is to survive and to produce young; and that those who are best suited to meet the changing demands of the environment, whether physically or behaviourally, are the most likely to be 'successful' in reproducing. 'Fitness' has a specific scientific meaning in this context. It is defined in terms of the number of reproductively viable offspring that you leave. Thus a male is not necessarily fit just because he is big and butch – he is fit only if this brings the females flocking to mate with him and if they produce young who are themselves fertile, and can pass on those beautiful butch genes. Natural selection is concerned with those survival mechanisms that can be passed on through the generations, those that are to some extent coded for in the genes. [7] Selection arguments usually focus on the individual as the unit of selection, but it is becoming increasingly fashionable to consider the individual itself as merely a 'gene-machine' (for a full-blown account of this approach see Richard Dawkins's *The Selfish Gene*). [8]

The biological basis of animal behaviour may be examined in two ways. At an anatomical level, scientists attempt to examine and describe the nervous system, and to relate their findings to the animal's behaviour. They look for the *cause* of the behaviour. It is comparatively easy to do this if we are referring to reflex actions, but it becomes far more difficult to understand the role of the nervous system in complex patterns of behaviour, such as are involved in an animal's social life.

The biological basis of behaviour might also be described in terms of the *functions* it serves. An evolutionary biologist looks at a particular pattern of behaviour, and asks, 'What use is that particular behaviour pattern to the animal in terms of survival?' Some patterns of behaviour, such as the crow of a cockerel, are relatively stable – that is, they appear to be very similar in nearly all individuals, and under most conditions. Other patterns appear to be more labile – that is, while they may be recognisable as belonging to that kind of animal, different individuals may show the pattern of behaviour slightly differently. For example, chaffinches all sing recognisably 'chaffinch' songs, but each individual chaffinch has his (females don't sing) own recognisable variant of the song. Evolutionary biologists have looked at the more stable patterns of behaviour and have asked whether, because they always appear to be so similar, they are largely inherited rather than learned.

The question of how much of an animal's behaviour is inherited is very difficult – if not almost impossible – to answer. We know that we can artificially select certain characteristics of behaviour when we selectively breed domestic animals – we have, for example, bred cattle that are fairly docile. Having done this for centuries, we know that at least some of the animal's behaviour is inheritable. The problem lies in attempting to define just how much can be inherited, and how much is subject to modification through learning. The study of sociobiology is simply the latest in the series of attempts to define how much of animal and human social behaviour is constrained by its evolutionary history; how much of it, in other words, is passed on from one generation to the next. It is important to bear in mind, in the following discussion, that the focus of a sociobiological approach to behaviour is generally on the limits to, rather than the potential for, social change.

The argument about the relative contributions of heredity and environment takes various forms. It appears as 'the venerable nature/nurture controversy', 'genes versus environment', 'instinct or learning', even 'fallen angel or naked ape?' Battalions of learned gentlefolk (and other less respectable folk) take up the old battle positions and shout the old war cries. One extreme calls, 'all that we think or do is determined at the moment of conception. We cannot change ourselves.' The other extreme shouts back, 'we are clay to be moulded in the hands of the environment. We are what we learn.' If this seems a touch too purple a piece of prose it is meant to be, for to take up either of these positions is clearly absurd. Inasmuch as we are made up of cells controlled by genes, all that we do has a genetic basis. There are no learning processes that do not have *some* genetic limits, but there are few, if any, behaviours that are dictated entirely by the genes.

Why then does the argument persist? It persists because the extremes are perpetuated not by the debaters themselves, but in their characterisations of each other. The trick is to accuse your opponent of taking up a patently ludicrous position, thus gaining for your side the extra kudos of demonstrated scientific rationality. You then move on to indicate that, while of course both genes and learning are involved, one or other plays a *predominant* part in the behaviour you are describing, thus arriving back almost where you started. For example, some 'vulgar' sociobiologists explain the poor response of women to 'equal opportunities' in terms of their overpowering evolutionary urge to be mothers and housewives. Wilson himself, in setting his argument up against 'the extreme orthodox view of environmentalism', plays a subtler version of this game.[9] Arguments about the relative contribution of heredity and environment in the shaping of human behaviour – and thus of human societies – are obviously of more than scientific importance. The Women's Liberation Movement has with good reason reacted strongly to the use of 'biological' arguments in the definition of the female; these have all too often been merely a confirmation of the patriarchal *status quo*. As a feminist and a biologist, I feel it necessary to state clearly my own position in the above debate.

I start with the premise that it is reasonable to expect that evolutionary biologists and students of animal behaviour will ask questions about the biological bases of human behaviour and societies. I see no reason to regard humans as other than biological beings, or to erect great barriers between humanity and the more 'primitive' forms of life. I consider this an insult to the rest of the animal kingdom; and in particular to our nearest evolutionary relatives, the apes. The richer we find their lives to be, the more it seems to me to indicate the failures of our simplistic scientific view of animal behaviour. Scientists have continually imposed a distorted view of the 'natural' on to animal societies, and then used the resulting 'evidence' from animals to demonstrate the 'natural' in humankind. It should be no threat to discover links between ourselves and other mammals – unless we believe that the science describes both us and them fully.

That said, I still believe that there is considerable validity in making a 'special case' for human culture in this context. 'Culture' implies far more than learning about our environment; human beings exist and act consciously within a woven complex of values, beliefs and norms which shape our conceptual world. Through the medium of spoken language culture provides an alternative means of passing on what we learn, which may bypass the genetic mode of inheritance. There is a far greater potential for change, but it is also true that culture exerts a powerful limiting force on individual action. Because biological arguments are usually framed within the context of the nature/nurture debate, the constraints that are imposed by cultural practices are often overlooked, and it is our biology that is seen as limiting our action. I want to look at the way in which such an argument is developed in sociobiological theory, in relation to the changing definition of women's 'natural' role. To do this, it is first necessary to go back to basics.

SEX AND THE SOCIOBIOLOGIST

Sex, it must be admitted, is a problem for the evolutionary biologist; sex means that you have to get your genes mixed up with someone else's. This is a hazardous business. Not only do you (theoretically) have to put twice as much effort into reproduction, as each of your children only carries half your genes,[10] but you run the risk of mating with a genetically undesirable type. Mixing genes does, however, have one great advantage – you might produce offspring with a 'better' mixture than your own. In a constantly changing environment a population in which the individuals are all slightly different will on average be better able to cope with change, and survive.[11]

The fact that sex exists is, however, all that need concern us here. In the mammalian species with which I shall be dealing, the problems of sex are seen to be of a different kind. Male and female approach the mating game with caution, and with different and conflicting aims. There is a basic asymmetry between male and female. The female knows that the young she bears are hers, and therefore worth her time and effort, while the male will never be quite sure. A female will be interested in two things – whether a male is carrying 'good' genes, and whether he is prepared to help her with the rearing of the young. The male however has different ideas. Because he is not limited by carrying and later nursing the young, his capacity for reproduction – known as his 'reproductive potential' – is far higher than the female's. His initial investment in the offspring is small, a teaspoon or so. As, in evolutionary terms, he is interested in getting the maximum number of children (thus passing on as many copies of his genes as possible) he will 'ideally' want to mate with as many females, and evade as much parental responsibility, as he can. He must also guard against cuckoldry, i.e., investing his time and energy in another male's genes. While in species where parental care is necessary for the survival of the young, it will not profit him to avoid parental responsibility altogether, as he is likely to be 'pulled' in the direction of desertion. Depending on the particular environmental circumstances of each species, widely different mating and parental systems may arise – from promiscuity to lifelong monogamy. In general however, according to sociobiological theory, 'it pays the male to be aggressive, hasty, fickle and undiscriminating. In theory at least it is more profitable for females to be coy, to hold back. . . .'[12]

In evolutionary terms then, the sexual relationship is 'one of mutual mistrust, and mutual exploitation'.[13] It is conceived of and phrased in the harsh terms of the market economy – investment and return, efficiency and maximisation of profits. Although the theory is of general application to animal societies, it contains within its formulation the (largely misogynist) seed of the rationalisation of many sexual stereotypes of Western society. If females must assess the genetic suitability of males, and their reliability as parents, then this may appear to provide an explanation for the 'feminine' qualities of coyness and grabbing possessiveness. If males are governed by a need to maximise their reproductive potential by mating with more than one female and/or ensuring the fidelity of their mate, then we might see as 'natural' polygamy, male philandery, male jealousy and violence against women. While these crude parallels exist only in the worst of the popular literature, they lie within the terminology of the science itself.

A LITTLE BROTHER SPECIES

The genetic basis of human social behaviour, which is a major concern of sociobiology, may be evaluated in several interconnecting ways:

1 by comparing us with our nearest evolutionary relatives, in order to look for similarities in social structure which might be accounted for in terms of shared genes;

2 by looking at the social behaviour of animals living in the environment in which humankind is thought to have evolved (J. H. Crook stated that 'under similar environmental conditions social animals will tend to develop similar social organisations'.[14] While this formulation is to some extent out of date, it remains part of the sociobiologist's parcel of assumptions);

3 by comparison across human cultures to look for cultural universals in human behaviour which might be the result of specifically human genes;

4 by looking at the most 'primitive' form of human culture, the 'gatherer–hunter' society;

5 by archaeological investigation of artefacts of human societies.

Although sociobiologists are increasingly moving into the field of anthropology, the major focus of their concern until recently has been a comparison between human and other primate societies.

> [A] picture of genetic determinism emerges most sharply when we compare selected major categories of animals with the human species. Certain general human traits are shared with a majority of the great apes and monkeys of Africa and Asia, which on grounds of anatomy and biochemistry are our closest relatives.[15]

In contrast to his earlier work, where he drew from a wide range of data on different primate societies, Wilson concentrates almost exclusively on the chimpanzee in *On Human Nature*. A quick glance at a range of books on primate societies will show that 'models' of chimp or other primate societies are fluid things – different areas of their lives are highlighted or played down, depending on the particular parallel with humankind that the author wishes to draw.

Since I am unlikely myself to avoid doing this, I am reluctant to add my 'reinterpretation' to the list. I do, however, think it worthwhile looking at the bias in Wilson's account. Wilson looks at several areas of overlap between our two species. He needs to do so in order to set up the possible evolutionary pre-conditions for those behaviour traits that he regards as our inheritance from a common primate ancestor.

As might be expected, the difference in size between male and female is considered to be of primary importance. Sexual dimorphism (differences) in body weight may be attributed either to different feeding habits of the two sexes or to sexual selection, members of one sex competing for access to the other. Although sex differences in feeding patterns are common in primates, [16] the latter explanation is more generally accepted in sociobiological theory. 'The theory of sexual selection states that the sex with the lower reproductive potential will be competed for by the other sex.'[17] In mammals it is the female whose reproductive potential is lower, limited by the time needed for the gestation and weaning of each batch of young. The male, theoretically at least, can inseminate many more females than the female requires males for insemination. Sexual selection has two components: males compete with each other for the resources of the environment – food, territory, females – and females select males. The more males have to compete for females, the more exaggerated they become – either in form, with elaborate plumage or colouration, in size or in behaviour, to 'advertise' themselves as suitable partners. Thus in primates there is a correlation between greater male size and the occurrence of polygamy. Polygamous species (one male with more than one female) show stronger sexual dimorphism, the bigger, stronger males gaining unequal access to the females while the weakest are barred from mating. Both humankind and chimpanzees show a small degree of sexual dimorphism (about 10 per cent), which is interpreted as indicating a trend towards 'mild polygamy'.

It is interesting to note that, 'although Darwin thought female choice an important evolutionary force, most writers since him have relegated it to a trivial role'.[18] It seems unlikely that scientists were far more ready to accept the notion of males fighting for possession of females than of females choosing their mates, merely owing to the lack of a theoretical framework in which to understand the latter.

An assumption underlying the theory of sexual selection as it related to males is that by reaching the 'top' of the dominance hierarchy some males gain unequal access to females and deny it to those males at the 'bottom'. In this context the female chimpanzee is interesting in that she seems to mate 'indiscriminately' as regards common evolutionary sense. She is notoriously promiscuous, and attracts many males during her oestrous period, without apparently incurring the wrath of the dominant male either for herself or for her partner. Such tolerance of mating with subordinates, despite the presence of a male dominance hierarchy, has also been recorded in other primates; some evidently distressed sociobiologists have pointed out that perhaps things are really still OK because the dominant males ejaculate more frequently.[19] Given that the object of the exercise is to pass on as many genes as possible, this seems to me to be a little too near Russian roulette to be a serious proposal.

The sex life of the female chimpanzee is not found in Wilson's account of chimpanzee behaviour. Rather, the female is subsumed within a male hierarchy (the male dominance hierarchy is seen as equivalent to male dominance over female), and her role within the society is largely ignored. This can be most clearly seen in the scant attention given to the role of the mother–infant bond within the group. In the early years of primatology the male dominance hierarchy was assumed to be the fundamental organising principle of primate societies:

> It is an unfortunate and now-recognised fact that many of the early field studies suffered from an over-emphasis on the behaviour of adult males. [More detailed studies began to show] another major axis of social organisation, the mother–infant bond, which ramifies through time into a mother-focused or matrifocal unit.[20]

While it is true that there has been a similar over-emphasis on the mother–infant relationship, in the wake of theories of maternal deprivation and the like, it does have major implications for the structuring of primate societies. The lengthening period of attachment of the young to their mother, which extends long after weaning and results in close links not only between mother and child but between siblings, has major consequences for the stability of the group over time. The matrifocal core provides a large network of social relationships over generations. It is at least as important as the male dominance hierarchy in the overall structuring of primate societies; and it is extraordinary that Wilson provides no discussion of it if, as he himself states, our long period of social training in infancy is an important and diagnostic behavioural trait.

For most of Wilson's account it is a case of 'cherchez la femme'. The society of the chimpanzee (which is used to provide us with a model of our primate past) undergoes a subtle transformation whereby the chimp emerges as primarily male, with the female's role largely ignored. The hunting male chimpanzee receives a large section of the text (despite the fact that meat forms an insignificant proportion of the diet), while the dominant female of the matrifocal core never appears. Wilson writes that his summary of the life of the chimpanzee is 'meant to establish what I regard as a fundamental point about the human condition, that by conventional evolutionary measures and the principal criteria of psychology we are not alone, we have a little brother species'. [21] Once again, it seems that our 'little sisters' are to join us in our obscure fate.

Wilson's selective treatment of chimp society prepares the ground for a similar over-emphasis on the human (male) hunter. A charge of bias cannot be laid at Wilson's feet alone. 'Man the Hunter' as prime mover in human evolution, both 'natural' and 'cultural', is an all-too-common theme. Of the theories of humankind's evolution away from our primate past, 'the hunting hypothesis' is still the most popular and the most popularised – reaching its lowest point in Robert Ardrey's book of the same name. Despite consistent attacks on Ardrey's vision of 'man the killer', [22] both anthropological and biological books perpetuate the view that the start of hunting (the evolution in men of weapon-making and handling, of co-operation and sophisticated methods of communication, etc.) provided the major spur to human evolution. For example:

> Hunting is the master behaviour pattern of the human species. It is the organising activity which integrated the morphological, physiological, genetic and intellectual aspects of the individual human organisms and of the population who compose our single species. [23]

It is tiresome, not to say depressing, to find brand new textbooks spending literally chapters eulogising the refinements of hunting behaviour, while devoting little or no space to the activities of the other half of the society. Supposedly enlightened talk of 'humankind' rather than of 'mankind' is of little comfort if the whole book then serves to support the time-honoured view of 'mankind evolving, womankind tagging on behind'. Since Wilson is not an anthropologist, he must rely on anthropological data which have usually passed through a filter of male bias and male expectation:

> The tendency to equate 'man', 'human' and 'male'; to look at culture almost entirely from a male point of view; to search for examples of the behaviour of males and assume this is sufficient for explanation, ignoring almost totally the female half of the species; and to filter this male bias through the 'ideal' western pattern of one male supporting a dependent wife and minor children. [24]

GATHERER–HUNTER LIFE

From the scant archaeological data that we have, it seems that humankind's first and most durable form of social organisation was that of the gatherer–hunter; a form of society based on a sexual division of labour, which is still present and much studied today. It is central to Wilson's thesis that present-day gatherer–hunter groups can offer clues to the genetic basis of human cultural behaviour:

> Human sociobiology can be most directly tested in studies of hunter–gatherer life. . . . We can be fairly certain that most of the genetic evolution of human behaviour occurred over the five million years prior to civilisation, when the species consisted of sparse, relatively immobile populations of hunter–gatherers. [25]

This argument requires that we make several assumptions. Of these, the first and most important is that 'culture arose because culture enhanced the genetic fitness of those who made maximum use of it'. [26] In other words, human culture has been shaped by natural selection in the same way as has human anatomy. The further argument, that present-day gatherer–hunters show essentially 'prehistoric' cultural behaviour (or '*basic* social responses', as Wilson puts it), requires that we assume that such groups are still subject to the same selection pressures as their (and our) forebears; that the original 'solutions' were genetically rather than culturally determined; and thus that social patterns have been preserved relatively unchanged. All these are assumptions rather than fact. The concept of 'subsistence' life in primitive societies, for example (with its underlying assumption of a struggle for survival which would forbid any wasteful social experimentation), has been questioned by the anthropologist Marshall Sahlins. [27] He has argued that people in these societies remain at 'subsistence' level because they work only long enough to satisfy their basic needs, and not because they are up against any environmental limits. [28]

The cultural forms of the gatherer–hunter societies may be as far removed from any genetic basis as our own. Wilson writes that 'sociobiological theory can be obeyed by purely cultural behaviour as well as by genetically constrained behaviour'. Faced with this challenge, the task of the sociobiologist in uncovering the genetic basis of 'human nature' is a heroic one: 'The possibility that the effort might fail conveys to more adventurous biologists a not unpleasant whiff of grapeshot, a crackle of thin ice.' [29]

Woman the gatherer

Gatherer–hunter societies are organised around a sexual division of labour where, very crudely, women gather and men hunt: an even cruder account would have it that 'men hunt, women stay at home'.[30] Some division of labour on grounds of sex appears universal enough to assume reasonably that it is very ancient; that it is based on inherited differences between the sexes is much less certain. Certainly we know nothing about how such a division first appeared, or under what conditions – or even what the original labour division was. In evolutionary theory it is argued that the pattern evolved to allow a more effective use of the food resources of the new savannah (open grassland) environment into which early humankind moved.[31] When meat was scarce, gathered or stored vegetable food could be eaten; when vegetation failed, meat was available. Women did the gathering because they could combine the work with child care.

The successful exploitation of such a division of labour required that male and female enter into a co-operative and mutually dependent relationship, which became extended far beyond the activities of food-gathering and -sharing. According to Wilson, 'many of the peculiar details of human sexual behaviour and domestic life flow easily from this basic division of labour'[32] – details of female sexuality, 'pair-bonding', child care, family life and the subjection of women. Evolutionary reasoning in this area too often starts with the assumption that man is the provider, woman the dependant. Woman the gatherer, as provider in her own right, is only slowly emerging from anthropological obscurity. Sally Slocum, a feminist anthropologist, even had to admit that it was only after she had become 'politically aware of myself as a woman' that she could ask the simple question, 'what were the females doing while the males were out hunting?'[33] The answer must clearly be: she was certainly *not* 'staying at home'. Women as gatherers play an important part in the economy of their society. Wilson quotes Richard B. Lee's study of sixty-eight societies where on average only about one-third of the diet consists of fresh meat. About 65 per cent of food for the community may thus be provided by the women.[34] Of course it all depends on how you put it:

> The male contribution to the subsistence base in hunting societies is always substantial, ranging from a little less than half in the tropics to nearly 100% in the arctic: the female contribution ranges from a negligible amount in the arctic region to a sustantial amount, but nowhere near 100%, in the tropics.[35]

'Gatherer–hunter societies are organised around a sexual division of labour where, very crudely, women gather and men hunt: an even cruder account would have it that "men hunt, women stay at home".'

A value-free assessment by a woman anthropologist!

Gross figures of the relative amounts of either meat or vegetables in the diet may obscure the real contributions of the two sexes, since female and male do cross the labour boundaries. Women may catch game themselves, or join in the hunt with the men, acting as beaters, as in the Pygmies.[36] Men may share the gathering, as for example in the Tiwi tribe of North Australia.[37] In these cases a division might be more usefully drawn between the pregnant and nursing women with their young children, and the rest of the group – a division that has also been suggested for analysing chimp societies where unhampered females join male groups.

While it is true that hunting that involves going after large game over long distances is an exclusively male preserve in gatherer–hunter societies, women are certainly not 'tied' to the home in the sense that we would understand it. Women's foraging activities may require an equal degree of mobility as does hunting. In the !Kung tribe (of the Kalahari region of South West Africa), for example, the adult women are absent from the home as much as the men.

Each adult woman goes out to gather about 2 or 3 days a week; on those days she walks from 2 to 12 miles, round trip, and on the way back carries loads of 15 to 33 pounds of vegetable foods. . . . More importantly, however, she also carries any of her children who are under 4 years old; those under 2 at all times, those between 2 and 4 part of the time.[38]

Women adjust their work pattern to the responsibilities of child care; those who are pregnant or nursing remain closer to the home base, while childless women or those whose children are sufficiently independent are far more mobile. Thus women's movement is limited biologically not by some inherited difference in exploratory behaviour, as has been suggested in the psychological literature, but because the biological events of childbearing and nursing place practical limits on them. Without these restrictions women are free to adopt other roles, and to cross sex-role boundaries (as are men) – unless they are restricted by cultural codes.[39]

Despite evidence of the flexibility of the labour division in many of these societies, it is more common to find evolutionary 'explanations' for sex differences that argue from the premise that the division of labour was and is based on pre-existing sex differences in behaviour. Thus, men became hunters because males were already 'stronger, more dominant, aggressive and powerful than women. They have greater physical endurance, are less fearful, more investigatory, and hence more likely to venture into and explore unfamiliar environments.'[40] Women became gatherers by default – because of what they were *not*, rather than because of what they were. Selection, so the argument goes, acted on women in one area of their existence, as mothers rather than as workers. If it acted on them as workers, it defined their ability to do boring and repetitive work.

Data from cross-cultural studies have however repeatedly defeated attempts to see in the cultural division of labour by sex 'evidence' for immutable sex differences in ability. The actual work assigned to the different sexes varies too widely across cultures for this to be true – what is women's work in one society is men's work in another. (Similarly, the attributes given to either sex vary widely.) It is evident, however, that the dual role of woman as provider and as childbearer has always been important in defining her work. A biological approach will always subsume the former to the latter. After all, 'the quintessential female is an individual specialised for making eggs'[41] (whereas the perfect male – human, dominant – is, according to Robin Fox, 'controlled, cunning, attractive to the ladies, good with the children, relaxed, tough, eloquent, skilful, knowledgeable and proficient in self defense and hunting'.[42] It must be the aftershave!).

Seeing woman's work as dictated by her reproductive function is only one way of approaching the problem. Some (female) anthropologists have suggested that 'just as the subsistence role of women must accommodate to the demands of child care, so childbearing must accommodate to the economic role of women'.[43] Patterns of childbirth may be culturally regulated, for example by taboos on intercourse at certain times, so that women can contribute their labour to the society.

Food collection and food distribution

The ways in which meat and vegetables are distributed among group members are usually different. Women tend to share their produce with their immediate family, while the meat that men provide is shared among the wider community.[44] Often it is assumed that the method of collection is the same as that of distribution – that women gather individually while men hunt collectively. This is not true. Women co-operate extensively; they share their knowledge of the bush, teaching each other where and how to gather, and they bring back vital information about the movement of animals to the hunters.[45] Visions of an exclusive realm of male bonding, male co-operation and male organisation, as opposed to the solitary, sheltered and dependent world of the woman, can be maintained only by ignoring women's work in these societies. Such a vision, however, is all too common, especially in the popular literature. It has been created to 'explain', by equating the all-male hunting band with the boards of directors/doctors/priests/lawyers/politicians etc., the 'naturalness' of our present male-dominated society. Only males co-operate; only males know how![46]

Differences in patterns of food distribution, as opposed to its collection, may however be related to the differential distribution of power. Ernestine Friedl points out that, in most human societies, sharing is a highly regarded behaviour, which confers prestige on the giver. If men are usually the ones who share within the community, they therefore have access to prestige and power which women do not have. Despite the fact that females have always shared food with their children, sharing in this wider sense is seen as a peculiarly male discovery. Yet somewhere along the evolutionary line, women had to learn to share too – if only with their mates.[47] We do not know how food was originally distributed, any more than we know what the original division of labour was. It may be that the exclusivity of the present patterns of collection and distribution were later cultural adaptations. We should consider the possibility that women's 'restricted food-sharing' may have been so defined as to *deny* her access to social power.

The social relevance of patterns of food-sharing have transformed the meaning of food within cultures. We may link the notion of food-sharing as a means to power with the relative value assigned to the two different food types by cultures. For Wilson, and for most anthropologists, the high value of meat as opposed to vegetables is self-explanatory: 'Only about one-third of the diet consists of fresh meat. Even so, this food contains the richest, *most desired*, source of proteins and fats, and it usually confers the most prestige on its donors.'[48] (My italics.) The question as to why meat should be more desired than vegetables – why something that forms on average only 35 per cent of the diet conveys to the donor higher prestige than that which may form 65 per cent, is thus answered solely in terms of the nutritional content of each. It seems unlikely however that the high value of meat is based solely on its high protein content – it is also possible that there is a two-way process involved, that the meat is also perceived of as valuable because of who the donors are. Maybe it is because men and not women provide most of the meat that meat is more valuable than vegetables.

If it were merely a fact of meat versus vegetables, then the point would be at best a trivial one, but it isn't. While the actual work that men and women do varies widely across cultures, men's work seems to be very often more highly valued than women's. If men hunt, then hunting has high status; if they build the shelters, that has high status – whereas if women do these things they are merely 'women's work'.

As in our society, the degree to which women are 'free' within gatherer–hunter societies depends on the degree of economic independence that they have. It is also obvious from anthropological studies that economic control is not a *sufficient* condition for freedom – power in a society rests on the value systems and reward structures of that culture. Women may make a major economic contribution and yet still be excluded from the areas of power that dictate the conditions of their lives. Judith Brown lists four dimensions that define women's status in subsistence societies: (1) female control over produce; (2) external or internal demand or value placed on female produce; (3) female participation in at least some political activities; (4) female solidarity groups devoted to female political or economic interests.[49] These dimensions might equally apply now, in industrial society. Given women's lack of political power, as women have moved into any area of employment, so its value within society has declined. Doctors in Russia are mainly female, and it is a low-status job. Secretarial work in the West was once a high-status male preserve.

The equivalence of 'male' and 'valued' seems to be almost as universal as the division of labour itself. For women this raises enormous problems: how do we begin to explain *why* human beings have the values that they do? And why, within so many of these value systems, is 'male' better than 'female'? Current sociobiological thought has little to offer us because it merely assumes the existing structure of our value systems. It may appear that, by seeing male dominance as an inherited primate behaviour, the higher value of things masculine might be 'explained', but it seems to me that the problem is in fact either unrecognised or ignored. The failure to recognise the importance of the value systems within which people operate in any society is for me clearly demonstrated in Wilson's account of the incest taboo and of 'hypergamy' – marrying up the social scale.

Incest taboo and hypergamy

The practices of the incest taboo and of hypergamy are used in *On Human Nature* as a testing ground for the 'infant discipline of human sociobiology'.[50] The job of the sociobiologist, when faced with such an apparently universal behaviour as ritualised incest avoidance, is to look for an underlying biological explanation for its presence. Incest avoidance was at one time considered to be a peculiarly human behaviour, but it now appears that it is widespread in both animal and plant kingdoms. Breeding with very close relatives brings heavy genetic penalties. The closer the relatives, the more likely they are to share defective genes, which, though masked in themselves, may be expressed in the joint offspring. As Wilson rightly points out, it is not necessary to understand the laws of genetics to observe the effects of inbreeding on your family. The disadvantages of inbreeding have to be weighed against the advantages of keeping the genes, and the money, in the family – marrying a cousin might make genetic as well as financial sense.

So far, so good – but can the sociobiological 'explanation' be regarded as a substitute for the earlier anthropological one? In that explanation the actual *form* of the incest taboo is seen as the most important factor; as Wilson summarises it, '[the incest taboo] facilitates the exchange of women during bargaining between social groups. Sisters and daughters, in this view, are used not for mating but to gain power.'[51] Not only 'in this view', but in fact. Women may be literally used as barter to cement social relations between groups of men. The two explanations of the incest taboo do not have to be regarded as mutually exclusive – while the sociobiological one identifies the 'deeper, more urgent cause, the heavier physiological penalty imposed by inbreeding',[52] the anthropological one interprets its cultural expression. Wilson himself identifies the trading of women, with its resultant social ties, as a 'secondary cultural adaptation'. Yet, when he moves on to postulate a biological basis for the practice of hypergamy, he fails completely to take this into account. Although the two customs are clearly linked culturally they are treated as isolated phenomena, and hypergamy is presented as an evolutionary strategy in its own right.

Women, according to Wilson, are 'beneficiaries' of hypergamy – by marrying men above them on the social scale, they gain access to 'better' genes. At the top of the hierarchy however it is of no advantage to be female, and women become the victims (beneficiaries?) of infanticide. In a section that combines naive anthropology and casual extrapolation from animal to human studies, Wilson treats us to an explanation of why the female children of high-ranking parents may be killed: because male children, with their higher reproductive potential, are more genetically profitable than females. From the same argument females should be more highly valued at the lower end of the social scale (since females are always in demand, and therefore mate, while low-rank males may not). There is however a strange silence here – perhaps because there is no anthropological data to support it? The argument for hypergamy as an evolutionary strategy should rest on an assumption of female choice, but certainly in the societies on which Wilson draws – pre-revolutionary China and caste-divided India – women have no independent control over their choice of partner. The problem of 'female choice' has attracted the attention of other sociobiologists; Richard Dawkins runs into similar trouble when trying to equate our modern patterns of dress with the plumage or colouration of other species. In *The Selfish Gene* he writes: 'It is strongly to be expected on evolutionary grounds that where the sexes differ, it should be the males who advertise and the females who are drab.' Why then, he asks, do we find the reverse in our society? Why are our females adorned and our males 'drab'? 'Faced with these facts a biologist would be forced to suspect that he was looking at a society in which females compete for males, rather than vice versa.'[53] Apart from the obvious flaw in the argument of limiting the discussion to the dress of twentieth-century Western society, it is clear that the dilemma is one of this biologist's own making, in trying to find a 'satisfactory' sociobiological reason for all and sundry. If he looked up from his sociobiological rule book, he might notice that, while our biology might still dictate that women should be the choosers, our cultural history has turned things on their head. In most societies women have become the property of men, who compete with each other *through* rather than *for* females. Even in our society it is debatable whether women have much choice, since their choice is still defined by a patriarchal history.

SOCIOBIOLOGY AND SEX DIFFERENCES

It is difficult to make an accurate assessment of the relative status of women and men in gatherer–hunter societies. Certainly gatherer–hunter groups such as the Pygmies or the Washo tribe of North America are among the most egalitarian societies that we know. Women and men have different spheres of existence, but where women make a major economic contribution to their society they generally have higher social status. Perhaps the most accurate reflection of this is the degree to which either sex will do the work of the other. In the semi-nomadic !Kung tribes, for example, there is a clear allocation of work on the basis of sex, but the boundaries are not rigidly observed; women and men can share both work and social power. Evolutionary theorists, however, tend to stress the separateness of the two sexes, for they argue that the 'biological basis' of reported sex differences rests in the different adaptations of the sexes to the demands of either gathering or hunting.

The study of sex differences is the subject of another chapter in this book. However I want here to look at some applications of Darwinian theory to the problems of the social roles of women and men, in the light of the division of labour. To some extent the debate has come full circle. Victorian scientists and social theorists argued for the equal status of women and men in 'primitive' cultures, but their purpose in doing so, and the conclusions that they drew, were very different from mine. The publication of Darwin's *Origin of Species*,[54] in which he put forward his theory of natural selection, had an enormous effect on Victorian society, quite apart from its purely scientific impact. The concepts of evolution and selection were indiscriminately applied to social systems. The hierarchical structure of society was seen to be a result of a process of social evolution – those at the top were there because they were most highly evolved. On a broader scale, Western society was thought to be more highly evolved than the 'primitives'; one mark of this lower state of evolution in primitive tribes was, according to Victorian concepts, the relative equality of the sexes. Women were seen to be so obviously more animal than men – in having menstrual cycles and giving birth to children – that societies in which women were considered to be equal to men must be less evolved. Theories about women were of two sorts: either they were considered to be 'evolutionarily retarded', remaining at the level of the primitive; or, if they had evolved it was as a parasite of man. Sociologists such as Durkheim in the late nineteenth century speculated that, as civilisation advanced, women became weaker and their brains became smaller. 'They depended more and more on men, and thus the family became a more stable institution. In short, as women declined, civilisation flourished.'[55]

Today the differences rather than similarities between women and men in 'primitive' cultures are stressed. It is argued that there has been a gradual divergence at the genetic level between the sexes, as each sex became better adapted to its designated sphere. Acceptance of this division of labour has apparently also been programmed in. Lionel Tiger writes:

> The contribution of non-maternal female behaviour to the genetic pool would be less than the contribution to the pool of those females who accepted a clear-cut sexual difference and enhanced the group's survival chances chiefly by full-time maternal and gathering behaviour.[56]

Wilson's own position on the issue of sex differences cannot be regarded as so determinist (in the sense that the differences are seen as rigidly fixed and unchangeable) as many of his predecessors, such as Tiger and Anthony Storr. It might be better to classify him as a 'wishy-washy liberal' (and I've been there myself so I mean no great disrespect!). One hallmark of this breed is the belief that, in order that effective (and cost-effective) social policies may be implemented, the differences between the sexes must be fully and *scientifically* understood. It is in such a light that Wilson approaches the question of the innateness of sex differences in humankind.[57] He first reviews some of the evidence for such differences, and then goes on to discuss the implications for society.

Two specific areas are dealt with – the areas of physical and temperamental differences between the sexes. It is here that, if Wilson and many other writers are to be believed, the roots of our oppression lie, for 'the physical and temperamental differences between men and women have been amplified by culture into universal male dominance'.[58]

The question of physical differences does not concern us here, for few I hope would argue for the physical equality of the sexes. It does however throw up a strange line in arguments; for instance, 'Women match or surpass men in a few sports, and *these are among the ones furthest removed from the primitive techniques of hunting and aggression.*'[59] Since Wilson's list of such sports includes precision archery and small-bore rifle shooting, one must presume that accuracy was not an evolutionary imperative in hunting!

The real debate about sex differences deals with the question of behavioural rather than physical traits. The 'masculine' traits of greater aggression, exploration, assertiveness and co-operation have become the routes to cultural power and control in our society. The evidence that these traits are biologically (by which 'genetically' is usually implied) determined rather than socially learned is often unconvincing. Certainly this is true of the evidence that Wilson musters in his somewhat conventional summary. He bases his argument on evidence from two sources: from studies of hormonally 'masculinised' children,[60] and on the supposedly asexual upbringing of children in the !Kung tribe.

It is doubtful whether studies of sexually 'abnormal' children, who have been repeatedly subjected to clinical examination, and whose parents bring them up in the knowledge of their ambivalent status, can be regarded as offering reliable evidence for biological rather than cultural determination of sex-specific behaviour. The 'evidence' from the !Kung tribe is even less convincing. That the 'asexual' upbringing of !Kung children (girls and boys apparently are not treated differently until puberty) still results in a sexual division in adulthood is not necessarily a testament to biology – it would be foolish to think that the children could not recognise and identify with those adult divisions. Even if they are not directly channelled into sex roles in childhood, they do not live in a sexless world.

Wilson's thesis does not rest on an assumption of absolute differences between the sexes, but on average differences between groups. He asserts not that men are aggressive and women are not, but that men on average have a greater potential for dominant behaviour than women, and that this tendency finds expression in almost all cultures. Now there is nothing really new about this argument, nor is it strange to women who daily come up against the effects of male dominance and aggression. If this is really all that Wilson himself is saying at a biological level, why am I wasting your time and mine in a seemingly nitpicking analysis of his latest views of women's roles? And what is potentially threatening to women about sociobiological theory? I want to draw together the threads of

Wilson's argument which I have gone into in some detail above, and to try and explain where it is that I part company with a socio-biological analysis of human behaviour.

A brief summary of Wilson's position here might be that one conservative trait of primate societies is the dominance of males over females. Males are generally more aggressive, assertive, etc., as well as being larger and stronger. Early human societies were organised round a sexual division of labour, which built on and reinforced the physical and behavioural differences between females and males. These have been culturally expressed in, for example, exclusively male organisations, the later elaboration of which has led to the subordination of women (as a result of cultural rather than biological evolution). If there is no necessary link between the division of labour and the subordination of women, there is a strong sense of inevitability in Wilson's discussion about the pattern of cultural evolution, given the 'biological' bias. More importantly, a sociobiological analysis assumes, or looks for, an adaptive function for social behavioural 'universals' – it must therefore assume that the structuring of human societies to give women and men different roles in society is to their advantage. For if it were reproductively dis-advantageous to women to mate with men who dominated them, they would have been selected during evolution not to do so. To follow this line of argument, women must be 'programmed' to support the basic system of male dominance, if not its cultural elaborations, much as men must be 'programmed' to be dominant. In its crudest form the argument all too often goes that women need/require/like to be dominated. There are real problems here in analysing exactly what Wilson himself is saying, since his work arises from within a long tradition of pseudo-scientific analyses of human evolution in which such crude arguments are put forward, and it cannot avoid being absorbed into that tradition. We are frequently warned not to put words into sociobiologists' mouths when criticising them; as John Maynard Smith wrote in his review of *On Human Nature*, 'I hope that criticism will be of things Wilson does say, and not of things that he does not.'[61] Wilson himself issues a strong caveat in his book against the use of his work for racist doctrines, but there is no such caveat against its use in sexist ones. Unfortunately, I think it unlikely that the message of his account of human nature, and the biological basis of human sex roles, will be seen as other than an ultimately determinist one, and that enthusiastic supporters of a vulgar sociobiology will tout these superficial views in support of their attacks on 'women's libbers'.

There has always been a clear trend in the writings of biologists who concern themselves with the social biology of humankind to dissociate themselves from the possible social implications of their work – and to portray themselves as merely the 'fact'-gatherers. Louis Agassiz, for example, a leading biologist of Victorian America and a proponent of extreme racist views (asserting that the different races are separate species), wrote:

> Here we have to do only with the question of the origin of men; let the politicians, let those who feel themselves called upon to regulate society, see what they can do with the results. . . . We disclaim all connection with any question involving political matters. [62]

The quote could just as easily have come from some of today's biologists and psychologists.

Wilson makes a similar division between 'scientific fact' and social policy. For him any struggle for women's rights must base itself on the unavoidable *reality* of sex differences as determined by scientific investigation. In the knowledge of these any society must make one of three choices:

> (1) condition its members so as to exaggerate sexual differences in behaviour (the present pattern of most societies) . . . (2) train its members so as to eliminate all sexual differences in behaviour . . . (3) provide equal opportunities and access but take no further action. [63]

Again, there is little that is novel in these proposals, and I do not want to devote time to the first two. The third is interesting, because it is avowedly the present aim of our society, and 'biological' arguments have been particularly prevalent in debates as to why such aims have 'failed'. In 1975 Wilson wrote:

> In hunter–gatherer societies men hunt and women stay at home. This strong bias persists in most agricultural and industrial societies, and on that ground alone appears to have a genetic origin. No solid evidence exists as to when the division of labour appeared in man's ancestors or how resistant to change it might be during the continuing revolution for women's rights. My own guess is that the genetic basis is strong enough to cause a substantial division of labour even in the most free and most egalitarian of future societies. [64]

This theme occurs in similar form in *On Human Nature*; and is given substance by the citation of the classic example of how women in the 'egalitarian' *kibbutzim* of Israel are returning to traditional 'feminine' roles. This has been continually used as hard 'evidence' for the innate structuring of human behaviour which will defeat such social planning as fails to recognise the 'natural' divisions of our society.[65] It is argued, from such an example, that a knowledge of our social biology is essential, if we are to define the options open to us in reorganising our society. Too often, biology is invoked to support the view that it is not worth it, either emotionally or financially, to invest to any degree in the 'liberation' of women.

There are alternative answers to the problem of why women don't rush into the top-level management posts that lie so invitingly open to them (?); and those alternatives lie not in the limitations of our biology, but in our cultural heritage. Women in our society, if they are to move out into the 'outside world', have to enter a specifically masculine world, and have to walk a continual tightrope between preserving their womanness and acting as honorary men. Not only this, but they often have to maintain their domestic role. The supposedly 'egalitarian' structure of the original *kibbutzim* required women to work alongside men, but did nothing to alter the patriarchal structure of Jewish society – men did *not* move into areas of 'women's work'.[66]

For any behaviour to appear, there must be both the capacity to produce that behaviour, and a social acceptance of it; in human societies the determination of whether a behaviour is sanctioned, and to what degree it is expressed, is largely a matter of cultural practice. (Margaret Mead's classic paper on three tribes in New Guinea showed how the expression of violence varied enormously across cultures, according to the social sanctioning of aggressive behaviour.[67]) It seems that the 'masculine' dimension of behaviour, which is only one dimension in the structuring of non-human societies, has assumed an overwhelming importance in most human cultures. It is not our female biology that limits us, but a culture that is often not of our making, but that serves to restrict us, by imposing limits on what we may do, and putting a negative value on so much of what we do do. That culture does not allow us the (biological) freedom to be 'equal but different' – our 'different' is defined as inferior, and that is a cultural not a biological statement.

SO WHAT?

How important is the theory of sociobiology for us? It probably seems somewhat irrelevant in the light of more pressing questions. The political and economic forces that determine the scope of our freedom as women do not rely on such science in any *direct* sense.

Science does however play an important indirect role, in providing 'justifications' for those measures, rather than a reason for their implementation. Scientists do not 'chance upon' important social truths while pursuing knowledge for its own sake, and then rush along with their evidence to policy-makers and demand social change. Historically, at least, science has instead tended to provide such 'evidence' as supports the ideology of the ruling classes who make the political decisions. Crude sociobiological ideas have been, and I believe will continue to be, used to help us 'accept' the restriction of our social role. In times of growing unemployment there is usually a call to give the jobs to the 'real' workers (the men), and we are then told that Nature not society has shaped us as housewives and mothers. There is a clear correlation between the appearance of such biological theories and the varying need for female labour. The cycle is continuous – recent newspaper articles have created the mythology of the 'hairy-chested' female executive, whose hormones just can't cope with the pressures of top-level management (which were apparently designed with the male hormones in mind).[68] New innovations, particularly in the area of computer technology, are widely forecast as being likely to bring a sudden and drastic change in employment patterns. The real fight for the right to work will not be at the level of scientific inquiry; but we can be sure that sociobiological concepts, lifted out of their 'pure' scientific context and applied with little regard to the ifs and buts of scientific debate, will be called upon to define us, and those who use them will rely on the weight of 'scientific objectivity and rationality' in the justificatory process. However much we may reject that image of science, it is a politically powerful one. We cannot therefore ignore the scientific arguments of sociobiological theory – we have to be able to answer the scientists as well as the politicians.

Part Two

SCIENCE AND WOMEN'S BODIES

4

The Tyrannical Womb:
Menstruation and Menopause

Lynda Birke and Sandy Best

Beware the womb. Throughout the history of Western medicine, this organ has been considered dangerous, and guilty of misdeeds so that its owners were not to be trusted, but had to be controlled. In most human societies, past and present, the functions of the womb are associated with strong social taboos, reflecting men's fear and awe of women.[1] Menstruation itself is often associated with women's seclusion, since there is a widespread belief that contact with a menstruating woman has dire consequences for men: for example, death, loss of hair or vigour, or possession by evil spirits.

This chapter is concerned with scientific statements about women's menstruation and menstrual cycles. However, since a form of the menstrual taboo does exist in our own society, and has existed for many centuries, we have started by outlining some of the effects of these fears on medical beliefs about women in the past. We also discuss the taboo in our own society, and how it influences the kinds of questions asked by scientists and doctors about women's menstruation. We have particularly concerned ourselves with the psychiatric literature on the menstrual cycle and the menopause, which seems to reflect social stereotypes of how women are expected to be, rather than how we really are. We have not only used medical information: we have also drawn on images from myth because we feel that this is at least as relevant as the strictly 'scientific' treatises. Finally, we deal with the treatment of older women by scientific medicine. The theme that runs through all of this is a view of women as sick, and hence inferior, and at the same time of women as potentially dangerous, and therefore in need of control.

THE TYRANNY OF THE WOMB

Women's obvious cycles have, throughout recorded history, been given a variety of explanations. To Aristotle, menstruation was due to an excess of blood and/or the phases of the moon to which it seemingly corresponded. Another idea among physicians of ancient Greece was that the cycle was due to the womb moving from its usual site. This idea proved to be very influential and persisted for many centuries. All kinds of 'female' complaints and vagaries of mood were attributed to these moves. An association between unstable moods and the womb was originally made by the ancient Egyptians, and was later taken up by the Greeks,[2] who gave the name *hysteria* (from the Greek for womb) to a particular kind of mental instability. A thousand years later, the cure for a wandering womb was to coax it back into place by fumigating the vagina with sweet-smelling odours while inhaling noxious odours through the nose! The 'wandering womb' idea only began to fade during the Renaissance as a result of the doctors who began to carry out detailed anatomical dissections of human bodies.

Like the moon, the idea that the womb could move waxed and waned over several centuries; but it was not until the eighteenth century that some doctors began to feel that the ovaries might be important for female physiology and behaviour. Although it is only in the last thirty years or so that we have had any real idea of what ovaries do, nineteenth-century doctors thought that they had much to do with 'personality disorders' of many kinds, an attitude that has been aptly described as the 'psychology of the ovary'. It was thought that a woman's entire personality was dominated by her reproductive system, to such an extent that it would suffer if, for example, she so much as put in four hours' study daily: 'If she puts as much force into her brain education as a boy, the brain or the special apparatus [that is, the reproductive system] will suffer.'[3]

The natural functions of the reproductive system, notably menstruation and pregnancy, were treated as abnormal and as a 'sickness': 'We cannot too emphatically urge the importance of regarding these monthly returns as periods of ill-health . . . when the ordinary occupations are to be suspended or modified.'[4]

This supposed domination of woman by her reproductive organs resulted in attempts by the medical profession to develop means to control her, such as by surgery to remove her clitoris or her ovaries. She seemed to change with her monthly cycle, and even to exhibit enhanced sexual desire at certain times of her cycle, which was considered to be dangerous. Surgery was proposed as an answer: after all, 'certain women especially at the menstrual epoch are so overcome by the intensity of sexual desires and excitement that they practically lose self-control and all modesty';[5] furthermore, 'orgasm was disease, and cure was orgasm's destruction'.[6] Middle-class men, aided by the male medical profession, needed their women to be predictable and unchanging, to act as a buffer against the stress of the working world.[7] It was in an attempt to keep women that way that they were being – literally – chopped and changed.

Since this role as buffer to the hard-pressed male was more charac-teristic of middle- and upper-class women, it is not surprising to find that the surgery was nearly always carried out on them (usually at vast expense). But working-class women did not escape. They were used as guinea-pigs for male doctors to 'develop' the appropriate techniques to use on the patients who could pay them well. One poor black woman in the United States was subjected to thirty such ex-perimental operations on her vagina.[8]

In this century, Western scientific advances have led to some understanding of what the hormones from the ovaries are,[9] and what they do. This understanding of hormonal action, however, does not include the effects on behaviour that were claimed by the nineteenth-century castrators of women. After ovary removal, they claimed, her moral sense was 'elevated' and she became 'tractable, orderly, in-dustrious and cleanly'.[10] Few modern scientists would make those claims about hormonal effects in women. Nevertheless, the idea that a woman's personality is dominated by her reproductive system still exists, as we shall see. The difference is that now the treatment is not surgery, but a pill.

Before we turn to modern medicine we should briefly mention some of the age-old myths and taboos surrounding menstruation. The reason for including this is simply that these myths and taboos reflect a fear and awe of women's reproductive function which is found in nearly all human cultures. In our own society, this fear of menstruation has contributed to the need for women to be 'con-trolled' by surgery, or by drugs, and contributes to the kind of questions asked by science about women.

FEAR AND AWE: WOMEN'S BLOOD

The majority of human societies, including our own, have some kind of menstrual taboo – that is, that menstruating women are feared, or considered dangerous, and are expected to keep their menstruation a closely guarded secret.[11] There are a few exceptions, such as the Congo Pygmies, who are encouraged to sleep 'with the moon' in order to conceive, and who associate menstrual blood with life.[12]

The menstrual taboos have existed for millenia, and no one really knows how they started. One suggestion is that women originally felt themselves to be at the centre of a mystery based upon their monthly bleeding, which related them to the moon-goddess.[13] Eventually, however, men became increasingly frightened by this mystery and its inherent power, and they developed means of setting women aside during their menstruation, and referred to the bleeding as 'impurity'.[14] Paula Weideger comments, in her book, *Menstruation and Menopause*:

> The menstrual taboo exists as a method of protecting men from a danger they are sure is real . . . and it is a means of keeping the fear of menstruating women under control. . . . For a woman, the taboo acts as a constant confirmation of her negative self-image. It represents the source of the shame she feels about her body and her sexuality.[15]

In other words, the taboo both protects men and serves to make women devalue themselves.

The menstrual taboo for most women throughout the world means ensuring that men never witness the fatal poison: women's menstrual blood. Women are secluded into menstrual huts, or they use internal 'sanitary protection' and pretend they do not menstruate. Menstrual blood is supposed to have terrible powers. According to Pliny, in the first century AD, menstrual blood was a fatal poison capable of destroying insects and dulling razors. Whatever women touched during their menstruation might have to be destroyed (for example in Uganda), or the woman had to be purified (for example, by ritual bathing according to strict Jewish law).[16]

Whatever its origins, the menstrual taboo is alive and well and living in our society. Most of us keep our menstruation a closely guarded secret, and do not like to admit to people around us that we are menstruating. Paula Weideger explores the effect of the taboo on women's own attitudes to menstruation, and on men's attitudes too. She describes modern versions of Pliny, suggesting that menstrual blood has its dangers. In 1875, for example, the *American Journal of Obstetrics* published a paper arguing that menstrual blood was the source of male gonorrhoea.[17] In our lifetime we have heard girls told that they should not wash their hair during their period as it was 'dangerous'. The clearest illustration of the taboo is seen in the number of people who find the idea of any form of sexual activity during menstruation 'distasteful'. This may, of course, become rationalised: Adrienne Rich[18] describes one man who said that 'intercourse with a menstruating woman did not appal *him* – but that it resulted in irritation of "the" penis'.[19]

We do not wish to go into details about the extent of the taboo in our society, as that has been done elsewhere.[20] For our purposes, we should point out that a taboo that is so pervasive, affecting everyone, must inevitably affect those people who would claim to be pursuers of rational objectivity. When scientists study menstruation and the menstrual cycle they too are affected by the values of a society that still considers women's bleeding to be so bad that it is tolerable only if it is kept secret. The taboo affected science to such an extent that it is only very recently that questions have been asked at all about women's menstruation: before that it was simply a case of making assumptions. It is significant that it took science until the 1970s to find out something that women had known for centuries – that if they lived together, their menstrual cycles tended to synchronise.[21] And then the research was done by a woman. It is significant too that although there *is* a vast literature on 'premenstrual tension' and painful menstruation, it is virtually *all* in esoteric medical journals, safely out of the hands of most women.[22]

Menstruation – the periodic loss of blood from the womb as its lining breaks down – occurs only in the 'higher' primates (apes and certain monkeys). In other mammals, if they do not become pregnant the lining regresses but it does not bleed, as it does not build up large blood vessels as in women. Although in prehistoric times women probably had few actual cycles (because of pregnancy and prolonged breastfeeding), that they had any cyclic bleeding at all might seem to be a disadvantage. So why did menstrual bleeding ever evolve?

We do not know the answer to that, but nor does science, simply because the question is rarely asked. Evolution theorists are quick to provide reasons for certain aspects of human behaviour such as 'male dominance', but they are remarkably silent on why we have evolved a periodic loss of blood from the womb. True, they have speculated about certain aspects of the menstrual *cycle*: for example, scientists have speculated about women's 'continuous sexual receptivity'. Females of most other mammals, of course, are 'receptive' only when they are on heat.[23] It has been suggested that this 'continuous receptivity . . . was responsible for the evolution of primate and eventually human societies'.[24] But this says absolutely nothing about why we ever *bled* in the first place. We simply do not know why.

IS MENSTRUATION NEW?

Bleeding is not new in our evolutionary history; but bleeding *every* month is. It seems likely that for most of our evolutionary history women have had relatively few menstrual cycles. Having started to menstruate later than we do, a woman would have become pregnant fairly soon, and then would breastfeed for a long time. Breast-feeding, especially in societies that are less well fed than our own, can inhibit ovulation and prevent menstruation for some time. Eventually the periods would return, and then another pregnancy would follow soon after. Thus, women probably had far fewer actual cycles than we do now: we can expect thirty to thirty-five years of about twelve cycles a year between puberty and the menopause. One scientist has suggested that this new phenomenon of repeated cycles might have disadvantages. We are now being exposed to constantly changing hormone levels, for which our bodies may not be fully adapted and which *may* even be bad for us in ways that we cannot yet predict.[25]

However, throughout history menstrual cycles have been a recorded phenomenon, even if they were not as regular as they now are. Some doctors believe that most women would readily choose not to have periods[26] if it were possible for them to do so, perhaps by using the Pill for much longer than three weeks at a time. We doubt that, however. For many women – despite the taboos – not to have a period can be a source of anxiety, and many women are now wary of the risks attached to long-term use of the Pill. We can see one big advantage to having fewer periods, though: it would markedly reduce the consumption of tampons and towels, whose manufacturers are already reaping enormous profits at our expense!

CHANGING MINDS: VARIATIONS OF MOOD AND THE MENSTRUAL CYCLE

The menstrual cycle means considerably more to most women than just monthly bleeding. Women, and the moon, which has the same cycle-length, have long been associated with *change*, which in the case of women is thought to be undesirable and in need of control. The association between the ever-changing moon and women is one that is represented in the mythology of many cultures throughout the world:

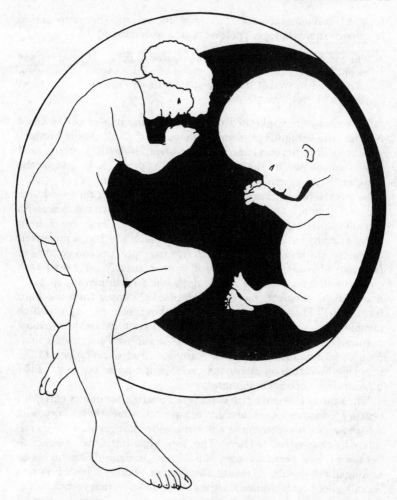

Figure Two: *'The Bambuti pygmies believe that women conceive when they "sleep with the moon", when they have sex at the time of the menstrual flow.'*

The lunar deity has been first and foremost related to the Virgin-Mother-Goddess who is 'for herself' and whose power radiates out from her maternal aspect . . . to the cycle of seasons, the dialogue of humankind and nature.[27]

It is a fundamental cycle, essential to nature, that appears so frequently in mythology. Yet it is fraught with danger:

> The sun is the constant and reliable source of light and heat, but the moon is changeable. . . . In these unaccountable qualities of the moon, man has seen a symbol of woman's nature which appears to him erratic, changeable, not to be relied on. [28]

This change is symbolised mythically by the images of the Good Mother, the benign face of womanhood, and the 'Terrible Mother, who holds fast, devours, and who is mistress of death mysteries' and who is symbolised by the terrible Indian goddess, Kali, and by the Gorgon. [29] The Terrible Mother terrifies man.

So woman has been associated in myths throughout the world with the changeable, unreliable moon: with darkness; with coldness; with terrible powers which bring about destruction. These associations pervade many religions: even in those in which the Sun is portrayed as female, she is changing and vain (for example, the sun goddess of Japanese Shintoism). [30] Not only is woman usually associated with change and darkness, but she is also associated in myth with madness: it is no accident that the word 'lunatic' derives from the word for moon. [31] To most peoples, throughout history, light (which favours the growth of harvest) and predictability (which favours planting or harvesting) have been highly valued; while unpredictability and changingness [32] are not. And today, much of the scientific literature on menstrual cycles continues to assure us of the undesirability of our changingness.

Since medical science first discovered ovarian hormones early this century, doctors have tended to assume that these hormones somehow directly determine the behaviour changes that they considered occurred in women. The idea that hormones determine behaviour is a pervasive one, but is not particularly useful when talking about people. [33] It is not always clear whether it is useful when we talk about other animals, although we do know that *some* kinds of behaviour in animals disappear if we remove certain sources of hormones.

In most mammals the female cycle is called an *oestrus* cycle, rather than a menstrual cycle. This word comes from the Greek *oistros*, meaning gadfly, indicating that the animal's behaviour changes as she comes into oestrus, or heat – as most owners of domestic animals will testify. Heat corresponds to the time when the egg, or ovum, is released from the ovary into the fallopian tube, where it may or may not be fertilised. Heat in other mammals, then, corresponds to mid-cycle in humans, half-way between periods, and does *not* correspond to our menstruation.

One very noticeable change occurring in other mammals at oestrus is that then, and virtually only then, will they allow a male to mate with them. In primates, and especially humans, one's sex life is less restricted. A woman's sex drive may vary with her menstrual cycle, but it does not switch on and off like that of other mammals. Although 'sex' is defined rather narrowly in biology as heterosexual intercourse (see Chapter 6 on scientific views of lesbians), it is still true to say that intercourse occurs in most other mammals *only* when they are on heat: in women this is obviously not true. The operation of the 'switch' in dogs, cows, horses and so on is under the control of hormones from the ovary, so that animals that have had their ovaries removed show very little sexual interest, unless they are given injections of hormones. In women, however, hormones may have a little effect on the sex drive (the Pill, for instance, may affect a woman's sex drive), but they are not *necessary* for her sex drive as they appear to be in most other mammals.

For most women, sexual arousal is possible at any cycle stage, although it is easiest around ovulation or menstruation. The *quality* of this arousal, however, may be different. It has been suggested that at ovulation women tend to be more likely to 'surrender' to their lovers; while premenstrually they are more likely to initiate sexual activity.[34] It has also been suggested that this increased sexiness at menstruation, especially since it tends to involve taking the initiative, can be 'disturbing to men reared on the idea that it is the male prerogative to initiate sex', which may contribute to men's fear of the menstruating woman.[35]

Most of what has been written about the menstrual cycle exists in the psychiatric/psychological literature, and refers to how women's behaviour changes. For many women, changes of mood during the cycle are a reality: depression or irritability, for example, may be felt just before a period. That these are very real does not, however, mean that they are directly *dependent* on hormones, as behavioural change seems to be in other mammals. We have already indicated how strong the menstrual taboo is in our own society: its existence, coupled with anticipation of discomfort for some women, may be sufficient to create symptoms, or to make symptoms far worse than they would otherwise be. There is some evidence to support this, since women from other cultures report different symptoms during the premenstrual and menstrual phases,[36] suggesting that the experience of menstruation and of cycles is shaped by the beliefs of the culture. We must stress that by suggesting this we are not doing what far too many doctors seem to do when they refer to 'psychosomatic' ills – that is, suggesting that they are imaginary. As women, we know only too well that feelings of illness, or being 'below par' are very real, and quite common. But we cannot ignore the possibility that the negative way our society views our menstruation may affect how we feel.

There are many changes in the body's functions, as well as in behaviour, that occur during the cycle. Brain waves (as measured by the electroencephalogram) are affected by the cycle, so that epileptic fits are least likely between ovulation and premenstruum, and most likely just before a period.[37] Various other functions change, such as carbohydrate metabolism (the rate at which sugars are used up by the body), thyroid function, mineral and water balance, resting temperature[38] and sensitivity to smells.[39] Many of these changes go unnoticed by the woman herself, however, whereas behavioural changes may be noticed by her or by her friends. It is significant to notice that it is mainly behavioural or perceptual (how you perceive the world) changes that are observed, either by women themselves or by men. Everyone has been trained to believe in the 'ideal' of remaining constant, so women are likely to notice any negative behavioural changes occurring within themselves. We all know that we are likely to become 'distractable', 'irritable', 'tearful' and so on premenstrually: we rarely hear about how we might become energetic and creative at mid-cycle.

In short, then, the changes that women tend to say they experience premenstrually are those that our society has deemed bad. Why should 'distractability' or 'tearfulness', or even temporarily putting on a bit of weight, necessarily be bad?

The psychiatric literature mirrors these 'bad' changes, and there are innumerable papers on the 'undesirable' changes occurring premenstrually. Many of these are based on studies of women using questionnaires that attempt to link certain mood changes with specific phases of the cycle.[40] The kinds of 'mood' examined by these questionnaires are, for example, decreased efficiency, dizziness, 'distractability', forgetfulness, lowered performance, excitement, and feelings of suffocation or nausea.

Most of these studies of the menstrual cycle have come in for criticism.[41] Generally, the methods employed in the studies leave much to be desired: questionnaires, for example, may be answered in a way a woman feels she *ought* to answer. Mary Brown Parlee asked women in one study to complete one of these questionnaires, and she also asked men to do the same[42] and to report on what they thought women felt. The answers were nearly identical, which led her to suggest that the 'premenstrual syndrome' of symptoms that are reported just before a period reflect what women (and men) felt was *likely* to occur. In other words, if there is a social consensus about 'what happens to women just before their periods' (as indeed there is: women are supposed to become bitchy and hard to live with), then women are much more likely to say that they feel those things.

It has been argued at various times that women's changingness, resulting from their hormonal changes, makes them unsuitable for certain jobs which require responsibility and mental effort. For example, one American doctor felt that women would not be suitable for certain positions because of their 'raging hormones':

> If you had an investment in a bank, you wouldn't want the president of the bank making a loan under those raging hormonal influences at that particular period. Suppose we had a President of the White House, a menopausal woman president, who had to make the decision of the Bay of Pigs, which was, of course, a bad one, or the Russian contretemps with Cuba at that time?[43]

In fact, there is some evidence that intellectual performance does not vary very much with the menstrual cycle. In studies of college women, for instance, intellectual performance was found not to change with the menstrual cycle, and college women apparently do not view their menstruation as particularly likely to disrupt their intellectual life.[44]

More to the point, the notion of women's unsuitability is a good example of the use of biological arguments to 'prove' women's inherent inferiority. Hormone theories are produced which imply that a woman's hormones render her better than men at boring repetitive tasks, while the converse is true of men.[45] You will rarely find ideas of hormonal determinism used to bolster notions of women's superiority at anything, with the exception of such tasks as child care. Furthermore, if women do succeed against all the odds in a man's world, then accusations of 'unfemininity' may be heard.[46]

Not surprisingly, there are many theories – too numerous to mention in detail – about the cause of premenstrual tension. Most of these theories involve the supposition that the woman concerned is suffering from a hormonal imbalance. There are, for instance, theories that premenstrual tension results from too little progesterone (one of the ovarian hormones);[47] or that adrenal hormones (from the adrenal glands above the kidneys) may be involved;[48] or that prolactin, the hormone primarily responsible for milk production after a baby is born, might be a factor.[49]

Although there are a number of hormone theories, we have some doubts about their validity. It may well be the case that *some* women have premenstrual problems that result, directly or indirectly, from certain hormonal imbalances; but hormone theories are less able to explain all instances of premenstrual problems. Such theories tend to imply that too much or too little of a particular hormone directly causes the psychological changes, thus ignoring all kinds of other effects. Most women, for example, know when their period is due, so there is always a problem of anticipation. In addition, the problems associated with the premenstruum vary from culture to culture, as we have already noted; and there is now mounting evidence of psychological factors (which we discuss below) contributing to premenstrual problems. There is much more to it than simply hormones.

Doctors carrying out clinical trials of new drugs usually allow for some psychological effects by conducting a double-blind trial. In such a trial, one group is given the drug and another group of patients is given an inactive sugar-pill. Neither the patients nor the doctors administering the drug know which is which. If only the group that had received the drug then improves, it can reasonably be assumed that the drug is effective and that there is no question of anticipation. On the basis of such trials, a few women do report that they feel better as a result of taking hormones.[50]

Nevertheless, there is evidence to suggest at least some psychological influence in the experience of premenstrual problems. When women take the contraceptive pill, the hormones produced by their own brains and ovaries are suppressed, so that ovulation does not occur. The contraceptive pill provides fairly constant hormone levels, with the exception of the week off, during which hormone levels decline and bleeding starts. As these levels are fairly constant, we might not expect premenstrual symptoms, if, as has been suggested, mood changes are directly due to hormone changes. However, in at least one study no differences have been found between women who took the Pill and those who did not: both reported that they felt worse premenstrually.[51]

Suggesting that there is a psychological component does not mean that hormones are not at all involved – just that they probably do not directly determine the behavioural changes. We know many women who have highly irregular cycles, and yet perceive some mood changes a day or two before their period arrives. This might be taken to support a hormone determination theory, but it need not. It is quite possible that, perhaps because of the negative way we value our menstruation, we become sensitive to slight changes in our bodies preceding menstruation. We might, for example, become aware of slight changes in our fluid balance. If we do, then the question of anticipating periods is still there, despite irregularity.

The chief significance of such hormonal theories is that they attribute the cause of a complex set of mood and behaviour changes[52] to women's biology. The reference to women's biology implies, first of all, that our social inferiority has biological roots and cannot easily be altered. More importantly, it serves to limit and restrict us, and to place the blame squarely on the individual. Rather than acknowledging that the society as a whole contributes to women's problems in the way that both women and menstruation are viewed, biological theories imply that the problem is within each and every woman. They thus serve to divert our attention from challenging the assumptions made by this society, towards wondering what might be wrong with *us*.

Much of the research on mood changes and the menstrual cycle focuses on what is seen as women's weaknesses: even the terms it uses are value-laden. In a society that favours reliability, obedience, energy, efficiency and even tempers, individuals who are believed to be showing periodic bouts of disobedience, unreliability and lack of efficiency are clearly less useful. Whatever changes are reported in such studies tend to be seen by the researchers, and indeed by the women themselves, in a negative light, as wholly undesirable. Rarely is any mention made of possible positive changes, such as feeling more energetic at mid-cycle.[53] It seems that the very fact of changing is seen as undesirable; that we are judged by the ideal of the supposedly constant male,[54] and found wanting. We might ask ourselves the question: to whom are these changes undesirable?

In the second half of the cycle many women feel more withdrawn, more 'in tune' with their subjective selves, a feeling that begins to change only as menstruation approaches. Then, near the time of the period, things begin to change again, and women are expected to become 'difficult to live with'. One book has suggested that the latter time is when the 'truth flares into consciousness'. Throughout most of the cycle a woman represses dissatisfactions, whether of her friends or lovers, or of the society that oppresses her. This repression fails at the premenstrual time, and the anger and hurt flare up: 'It is the "moment of truth", which in a society which refuses woman her true place may become the moment of despair.'[55]

The reader may find this idea exciting, or repellent: it does not much matter. It is certainly an interesting idea, and one that women may find helpful in understanding their own changes. It is surely time that we refused to accept scientists' negative evaluation of our cycles, and to learn to evaluate our changes positively. Altering this will not be easy, in terms either of scientific attitudes or of ourselves. Even in the context of the women's movement, there is remarkably little interest in, or knowledge of, the menstrual cycle. Women's health groups have developed techniques of menstrual extraction, which promises to get menstruation over quickly, but little has been developed to help us further our understanding of what menstruation means to us. Harding says:

> It requires an act of devotion deeper than at first seemed necessary, if a woman is to live her life in harmony with the rhythm of her own nature. . . . For the rhythmic life within her is determinant of her own life, while her conscious wishes and impulses do not necessarily coincide with her deepest needs.[56]

Finally, we must pose the question, ignored by the researchers: why is changingness so undesirable? The sun–male–constancy association within the mythology of so many cultures is imposed upon us. We are led to believe that men are constant and that this is good. Women are swayed by strange forces and are not to be trusted: within mythology they are moon-changing. We now live in a supposedly scientific age, when scientific logic and rationality, rather than superstition and mythology, are the standards by which we judge our world. Yet the beliefs illustrated in the mythology surrounding women's menstruation seem to be perpetuated in this 'scientific age'. The questions asked, the ways in which the results are formulated or interpreted, all contribute to a negative view of menstruation and to a view of woman as a being gripped by 'raging hormones'. Because of these, it is often said, she cannot be trusted or given responsibility: she should be kept in her place.

REWARD OR PUNISHMENT: CULTURE, BIOLOGY AND THE MENOPAUSE

We have said that it is meaningless to us to consider only the biological changes that characterise the menstrual cycle, without considering the socio-cultural factors that affect all women. The same can be said of the menopause. The ways in which a society treats the older woman determine to a large extent how she will adjust to the obvious changes occurring in her body's functions. Some societies, such as the high-caste Rajput classes of Rajasthan in India, reward women, in that the constraints normally imposed upon them during their fertile years (such as seclusion) are removed. Other societies tend to punish women who are menopausal. Our own society now lays strong emphasis on youth and beauty: the image conveyed by popular magazines is that youth is fun, energetic, productive, passionate and carefree. Older women are not particularly valued. The consequences of ageing can be traumatic for women: they may feel useless as their children leave home; they may become hopelessly depressed;[57] they may feel that they should have plastic surgery to correct facial lines or sagging breasts; they may begin to use cosmetics extensively.[58] As Paula Weideger comments:

> The menopausal woman has had her social education. Within its framework her soul is now of little importance because fertility and 'dangerous' menstruation no longer form part of her being. Her body is no longer unclean, but it no longer serves its most useful function – that of giving form to fertility. This body, however, 'lingers' on, and what remains becomes the focus of anxious attention.[59]

In centuries past, when the lifespan was shorter and the risks of both maternal and infant mortality very high, a concentration on women's fecundity was understandable. Yet in an increasingly populated world, and with increasing numbers of people surviving into old age, it becomes anachronistic. One might have hoped that scientific and medical books would not suffer to the same extent from this bias towards reproduction, but the hope is likely to be disappointed. One recently published book for example, about the physiology of sex hormones, concentrates almost entirely on fertility and contraception – how to induce ovulation in women who are not having periods and so on. The menopause is dealt with in approximately ten pages (2 per cent of the book), despite the extensive current use of hormones to treat menopausal symptoms.[60]

The menopause is the cessation of menstruation (another term often used is the climacteric, which refers to the passage from middle to old age). Slowly, various changes take place in the body. Some of these are due to general ageing processes, although they may be intensified by the declining hormone levels, such as bone-brittleness (osteoporosis), and changes in the skin cells. Other changes directly involve the reproductive system itself.

Medical science does not really know what happens during the menopause (see the Appendix for a brief outline of the biological events that occur), and research on it is still in its infancy as it was not considered a particularly important area to work on until recently. What is known is that somehow the link between the pituitary gland – which controls the output of most of the other hormone-producing glands – and the ovary breaks down. Certainly the ovary starts to produce smaller and smaller quantities of its hormones. It is the decline in oestrogen levels from the ovary that produces many of the effects of the menopause that women may find undesirable, such as 'hot flushes' or dryness of the vagina and vulva.

A woman may experience many symptoms as she goes through the menopause: she may get hot flushes as oestrogens decline; she may feel tired and dizzy; she may suffer from insomnia or depression; and, above all, her periods become more irregular, and eventually cease altogether as the link between pituitary and ovaries finally breaks down.[61] Dryness of the vagina can also cause distress, as well as an increase in vaginal infections.

These symptoms are real enough to the woman herself, although doctors believe that not all of them are definitely organic. Hot flushes and drying of the vagina are certainly facilitated by the decline in oestrogen levels, but some of the other symptoms do not occur in other cultures. This suggests that some of the severity of the symptoms may be due to how women feel about the menopause, and how they see themselves. The image of femininity in our culture is inextricably linked with youthful fertility, so it is hardly surprising that women look upon the menopause with fear and dread. Yet we can now expect to live on average a third of our lives beyond the menopause.

> Women who are sharing this experience now grew up in an era when the definition of the female role accepted by the majority of women – willingly or reluctantly – centred on fertility and motherhood. . . . As recently as the 1950s a woman who did not wear a wedding-band or push a baby carriage was hardly considered a woman at all.[62]

Many of these women who grew up in that era now turn to the medical profession for help: they experience a sense of futility and worthlessness, coupled with troubling physical symptoms. Some doctors dismiss them totally with comments such as 'it's just your age'; or, 'it's women's troubles'. An increasing number today are prescribing hormone treatments for the menopause.

Although oestrogens are not entirely absent in post-menopausal women, they are reduced by about 80 per cent. Synthetic oestrogens are often given to relieve the symptoms, and they can be given in various ways. They might, for example, be given at low doses on their own – which will prevent hot flushes – but in too low a dose to cause any bleeding from the womb. Alternatively, they might be given in conjunction with a progestin (similar to progesterone, which is a normal ovarian hormone) for three weeks followed by one week off, during which time the womb sheds its lining. On the latter regimen, women might feel that they have not 'lost' their periods. On the other hand, they may well experience hot flushes during the week off the pills.

Oestrogen treatment (more usually called hormone replacement therapy, or HRT) reverses many of the 'undesirable' effects of the menopause – at least as long as a woman goes on taking the pills. It eliminates hot flushes, reverses the drying of the vagina, and to a lesser extent inhibits bone-brittleness that accompanies ageing. It also combats 'masculinisation',[63] which accompanies ageing in some women, and results from hormones from the adrenal gland which predominate in the absence of large quantities of ovarian hormones.

There are many advocates of HRT, both among the medical profession and the public. The drug companies, of course, might be expected to support it: large profits can result if women take a pill daily for twenty years. Advocates among the medical profession make such cautious statements as: 'In weighing the benefits of oestrogen therapy mental, physical and metabolic, against undesirable side-effects and risks, the conclusion has to be reached that the benefits exceed the latter.'[64] Rather more extreme advocates can be found: in one popular book about the menopause the author states the belief that oestrogens are *essential* for femininity:

> For [oestrogen] is the key to a woman's health and happiness, and it is at work in the body even before birth [which is not strictly true]. . . . In many ways oestrogen has a place in the life of a woman rather like that of love itself. She is born with both, and both assume their greatest importance in her fertile years.[65]

The same author directs such spurious arguments towards the medical profession too: 'I believe', she says, 'it is the responsibility of medical science to ensure not just the quantity but the quality of this extra life [prolonged expectancy of life] with *proper hormonal support*'[66] (our emphasis).

The implication of such statements is that, if we continue to take these drugs, we can remain feminine forever.

As we have indicated, many of the effects of oestrogen therapy appear to be partly psychological – even hot flushes respond to some extent to placebos (sugar-pills).[67] One report suggests that, of all the symptoms, only insomnia appears to be solely due to oestrogen 'deficiency'.[68] In view of this, too much enthusiasm about the wonders of HRT is premature. Quite simply, there has not yet been enough research to indicate whether taking drugs for long periods of time might be hazardous to health.[69] There is controversy, for example, over whether certain forms of oestrogen might facilitate some types of cancer.[70]

The medical profession in this country may now be more cautious in prescribing artificial hormones to older women (other than for short-term use to counteract some of the more distressing menopausal problems while they are at their worst), in the light of recent reports that hormones given to older women increase risks to health.[71] Such caution is long overdue. Weideger comments: 'It is not enough to say that every woman has a right and a privilege to choose her poisons and name her saviors. I want to choose my medicines *after* I know their side-effects, not after their side-effects have ravaged my body.'[72]

By pushing HRT, doctors are little different from the rest of society which adopts a rather punitive attitude towards the older woman, when they say, with reference to the week off the drugs: 'It was . . . suggested that reappearance of hot flushes during the medication-free week could serve as a useful reinforcement for the patient, *reminding her how good the therapy really is for her and how badly she needs oestrogens*'[73] (our italics).

This hard sell of oestrogen can be compared with the attitude of Chinese doctors. Few doctors in China advocate treatment for the menopause, and symptoms seem to be rarer than they are here. When women do have severe symptoms, they can be given gentle herbal treatments. In China older people are respected, and have a useful role in the community: the menopause seems not to be the negative, shameful thing it is in our society.[74]

We started this chapter by referring to the menstrual taboo within our own society. By advocating HRT to simulate menstrual cycles for years on end, doctors are giving us the old double-edged sword of 'prolonged youthfulness' (with possible drug risks attached), *and* reaffirming the taboo. Women are persuaded how much they need the wonder drug by appeals to their desires to remain 'feminine'. For women whose self-evaluation has primarily been in terms of their 'femininity', their qualities as wife and mother, the changes of the menopause are indeed threatening. Many women need to be reassured that they are still as feminine, as desirable, as they were when they were young. But the answer is surely not likely to lie in daily doses of a pill that is sold with the promise that it might keep you feminine.[75]

In the nineteenth century attempts were made to control women by surgery. In the twentieth century we are controlled and denied our strengths by attributing to us alarming moods which must be controlled, and by giving us drugs accompanied by empty promises. How far have we come in 100 years? We still keep our bloody secrets; we still know and understand little about the menstrual cycle; we still fear the menopause. Scientific inquiry has so far contributed little to dispelling the prejudices and mysteries surrounding our monthly bleedings.

As we write this, we two are still young, still menstruating. One day, we will go through the menopause just as our mothers are doing now. We hope that we cope with it as well as they do.

5

From Zero to Infinity:
Scientific Views of Lesbians*

Lynda Birke

Dedicated to those women who have been punished for their lesbianism, in the name of a cure.

This chapter is about how science views lesbians. Within the last hundred years, numerous scientific theories have appeared, suggesting possible reasons for the 'aberrant' behaviour of lesbians, and of gay men. In a society that condemns homosexuality,[1] it is almost inevitable that scientists and doctors living in that society have absorbed those values to some extent; and so they ask questions about lesbianism not with the objective eye of ideal science, but more with assumptions based on centuries-old prejudice.

The first part of this chapter is about ideas of lesbianism, and the second part about medical 'treatments' that result from these ideas. Prejudice against homosexuals of either sex has a long history, and it is with the history that I shall begin. Since much of the scientific literature on homosexuality makes assumptions about the biological unnaturalness of homosexual behaviour in humans, I also briefly consider whether lesbians are found in other cultures, before looking at the ideas produced by science itself.

*This title was suggested by the statement of a doctor unsympathetic to lesbianism: 'One vagina plus another vagina still equals zero.' (Reuber, D., *Everything You Always Wanted to Know about Sex—But were Afraid to Ask*, London & New York, W. H. Allen, 1970, p. 215.

IDEAS ABOUT LESBIANS IN THE PAST

Homosexuality has for centuries been condemned. It was, and is, considered a heresy according to theological doctrine; alternatively, it has been thought of either as a sickness, or as a 'viciousness' which should be punished. Nearly all the references to homosexuality in the past have been to men; lesbians were invisible, at least as an idea. This is *still* a feature of nearly all writing about homosexuality. A recent book about homosexuality[2] deals mainly with gay men, while about four times as many articles appear about gay men as about gay women in medical journals.[3] Charlotte Wolff[4] suggests this apparent invisibility of lesbians is due to men seeing lesbianism as a joke, something not to be taken seriously. As a result, she says: 'It is the arrogance of the male which has protected homosexual women from suffering the same degree of persecution as homosexual men.' Many lesbians would not now agree with her that they are protected by male vanity: there have been far too many recent cases of assault on lesbians in our cities, simply because they are women daring to live without men.

Thomas Szasz, in his book *The Manufacture of Madness*, points out that for centuries no distinction was made between religious unorthodoxy and sexual misbehaviour, especially homosexuality.[5]

This association resulted in accusations of heresy for all kinds of sexual misbehaviour, real or imagined.

Burning witches at the stake was justified during the witch-hunts of the fourteenth and fifteenth centuries by claiming that these unfortunate women had 'inclined the minds of men to inordinate passion', and had somehow managed to make men's penises fall off as a result of intercourse with them; these could then be used for magical purposes![6] Women's sexuality has always been feared (and punished) by men, so it seems highly likely that lesbians were included in the witch-hunts: unfortunately, history tells us little.[7] Apart from the stake, however, there were other punishments: in the thirteenth century, the Legal School of Orléans specified punishments for lesbian offences. For the first offence, clitoridectomy (removal of the clitoris) was carried out; for the second, further mutilation (unspecified); and if a woman was so intransigent as to try it a third time, she was burned at the stake.[8] Genital mutilation has at times been seen (and indeed still is seen in certain parts of the world) as a means of controlling and containing women's feared sexuality: this thirteenth-century treatment of lesbian sexuality is certainly not unique in this respect.[9]

As homosexuality has so often been viewed as a sickness, it is not surprising that there have been pages of print devoted to speculating how it develops. Yet there are none speculating how heterosexuality develops; although Freud[10] pointed out that surely the 'exclusive interest felt by men for women is a problem that needs elucidation, and is not a self-evident fact'. No one, it seems, has subsequently felt the problem to be worth consideration. Heterosexuality is assumed to be 'normal', largely because it (potentially, at least) leads to reproduction. Sexuality has, in the Christian ethic, long been linked to reproduction. In the early years of this century, medics had lengthy arguments about the morality of contraception, as it was felt that it was wrong, and going against nature[11] – a position still officially maintained by the Catholic church. It was held to be immoral because (1) it did not allow the possibility of reproduction; and (2) it allowed the possibility of sex for sex's sake. To the church, *any* expression of sexuality that did not potentially lead to reproduction was sinful; similarly, to many people, homosexuality is wrong precisely because it *cannot* lead to reproduction.

The notion of the gay person as somehow sinful/evil has persisted to the present day. Gays are thought of as threatening, and in particular are thought of as likely to seduce or pollute children (hence the anti-gay campaign in the United States in the mid-1970s entitled 'Save Our Children'). It is commonly thought that gays should be debarred from professions involving contact with children, such as teaching, and certainly from parenthood.[12] It is also significant that it was not until the early 1970s that the American Psychiatric Association removed homosexuality from its list of mental illnesses, and re-classified it as a 'disturbance' (and even then the vote was far from unanimous: 5,800 to 3,800). It is certainly true that many people believe homosexuality (in either sex) to be unnatural: but just how 'unnatural' *is* homosexuality among human societies?

THE OCCURRENCE OF HOMOSEXUALITY

It is doubtful indeed whether our society would have such strong sanctions against loving someone of the same sex as oneself if such a love had *never* existed. Homosexual behaviour is certainly not rare in the Western world at the present time, as was indicated in the Kinsey report,[13] and more recently in the Hite report on women's sexuality.[14] Lesbian relationships are accepted and actively encouraged within the ethos of the Women's Liberation Movement at the present time.

Homosexual behaviour among men is accepted as quite *normal* in at least half of the societies studied by anthropologists.[15] There seem to be fewer societies condoning lesbianism, although some evidence is beginning to emerge: for example, it has recently been reported that lesbianism can be an accepted social choice among women living in Mombasa, Kenya.[16] That there are fewer reports of societies that accept lesbianism probably reflects the fact that most anthropologists are male, as well as being from a society that prefers not to think about homosexuality. Consequently, they may overlook lesbian relationships or not be told about them: women are, not surprisingly, reluctant to divulge secrets of their sexual/affectional lives to strange men. There may, of course, be fewer societies anyway in which lesbianism is accepted. When a society is living in a subsistence economy, for example, woman's reproductive role is highly valued as it provides further hands to work the land, take animals to market and so on, as well as providing heirs. In this situation, one might expect to find some social sanctions against lesbianism (at least, against exclusive lesbianism), as it would remove women from their economically important role as bearers of future labour.

Some human societies view homosexuality with disgust, while

others sanction it. Social sanction may even allow someone (although it is nearly always men) to step outside the boundaries of what is permissible behaviour for their sex. In Oman (in the Middle East), for example, some men can act as homosexual prostitutes, [17] yet they are treated, and behave, socially as though they are women – even to the extent of being allowed to gossip with the women who are in seclusion (purdah). For other men, living socially as men, it is forbidden to enter the area of women's seclusion or to see their faces (other than those of the immediate family). Later on, transsexual men can change back to the male role and get married, after which they live exclusively as men. Clearly sex roles, at least for Omani men, are not immutable.

Even from the few examples I have given here, it should be evident that homosexuality is not rare among human societies. It exists almost everywhere: what changes is how each society views it. There is certainly little enough evidence to suggest that homosexuality is biologically *un*natural in human societies. It is only unnatural if sexuality is inextricably linked to reproduction. Nevertheless, assumptions of 'abnormality' thread through the scientific literature about homosexuals in the same way that notions of 'sin' used to thread through the religious literature. A gay person may be described, for example, as having 'too much' or 'too little' sex drive; as 'over-identified' with her father/his mother; or as 'immature' and 'unable to enter into normal, mature relationships'. For women, of course, the latter usually means marital domesticity and subjection to husband and children.

Let us look first at some of the biological 'causes' of homosexuality which scientists have suggested, and the ideology on which they are based.

HORMONES AND THINGS

There are many biologically based theories or explanations of homosexuality (for example, the 'wrong' hormones; congenital abnormality, etc.), which are based on a number of unquestioned assumptions. Scientists apparently find it necessary to look for an explanation of homosexuality, but not heterosexuality: one is thought of as abnormal, and therefore needing to be explained; the other is felt to be so normal that explanation is rendered unnecessary.

The first assumption on which these theories rest is that heterosexuality is the only normal expression of love and sexuality in

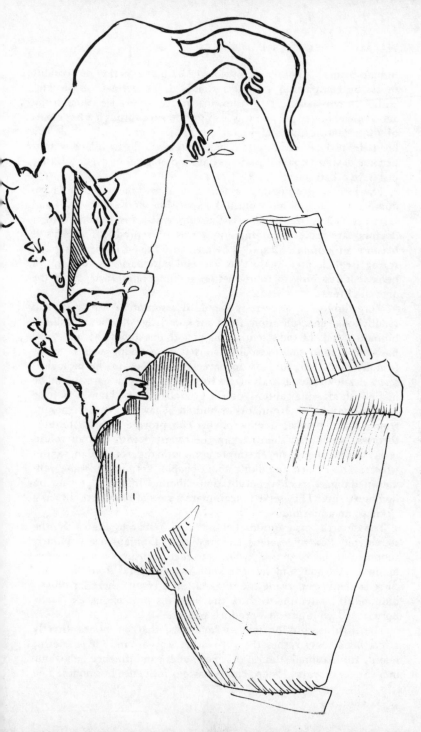

'The usefulness of studying what rats do in bed?'

human beings. All other expressions of human affection or sexuality must, by comparison with this standard, be defined as deviant. Although our society is now showing greater tolerance (though not acceptance) of gay people, this assumption runs through a large part of the scientific/medical literature. Once a group of people has been defined as deviant, it is not unreasonable to ask why they become deviant: could it perhaps have a biological basis, or did they just have a 'bad start in life'?

The second assumption is referred to above, namely that reproductive sex is the only normal expression of human love and sexuality. The problem here is that biologists, and to some extent doctors, are trained to think in terms of reproduction. This is because an animal's 'fitness' is defined in terms of its surviving offspring (see Chapter 3), so that we tend automatically to think of behaviour resulting in reproduction as 'functional' in terms of the animal's fitness.

More dubious, however, is the third assumption – that we can readily make generalisations from our knowledge of other animals to humans. Many of the scientists making claims about the biological basis of human homosexuality started by working with animals (usually rats). Most of us who have been trained in the biological sciences know that generalisations from animals to humans should be made with extreme caution, on the grounds that we know very little as yet about the determinants of human behaviour. At best, results from animal behaviour experiments can provide us with testable theories, which can then perhaps be re-tested with human volunteers; but we should never simply jump straight from talking about what rats do in 'bed' to talking about people. Yet, despite these well-known dangers, such generalisations abound in the literature on homosexuality. They serve to legitimate a view of homosexuality as a sickness, an abnormality.

The fourth assumption behind this research fits with the stereotypic view of lesbians and gay men. Lesbians are popularly supposed to be 'mannish', while gay men are supposed to be 'effeminate'. Although many of us know that this is far from the truth, these stereotypes pervade the research: as a result, there have been innumerable attempts to show that lesbians produce excess 'male' hormones[18] while gay men produce 'too little'.

The fifth assumption follows from this: that hormones directly affect behaviour, rather than the other way around. It is known, mainly from animal studies, that hormones can influence behaviour in a variety of ways – but so can behaviour influence hormones. The calls of many male birds, for example, attract the female and she engages in nest-building activities. These, and the male's courtship, bring about a rise in the female's hormone production, thus bringing her into reproductive condition. Similarly, stress is known to have a considerable effect on androgen secretion in adult men.[19] Although there is mounting evidence of behavioural effects on hormone secretion, it does not often percolate into the literature on homosexuality. Suppose, for example, that a study reported that a group of lesbians had higher levels of androgens than heterosexual women (as indeed some reports do). The commonest interpretation of this is in terms of the hormones influencing the lesbian behaviour. Much less commonly, the possibility is raised that the behaviour has influenced the hormones.[20] This, of course, need not imply that it is the lesbian's sexual behaviour that is crucial: there is a considerable difference between the life-style of a 'typical' lesbian and that of a 'typical' control. Any aspect of this life-style might, in theory, affect hormone levels.

There is, in fact, a third possible way of looking at such a finding, and it is one that is never entertained within the heterosexual norms of the literature. This would be to say that something about the heterosexual life-style of sexuality leads to a *reduction* in hormone levels relative to those of the lesbian group. But such explanations are never considered or tested experimentally: the standard is the hormone levels and behaviour of the heterosexual group, against which lesbians are judged.[21]

But what of the animal research on sex hormones, on which much of this is based? A crucial area of study is that of the effects of the sex hormones (that is, those produced by the ovary or testis, and also to some extent by the adrenal glands above the kidneys) before an animal is born. Scientists can study this by giving the hormones to animals before, or in some cases just after, an animal is born. For example, a new-born female rat may be given androgens ('male' hormones) and be more likely than her normal sisters to show some behaviours that are defined as 'male' when she grows up. She may mount other individuals more, for instance, and be less inclined to show passive receptivity (defined as a female response!). The hormone treatment has 'tipped the balance' of the brain, as it were. Normal female rats do in fact mount others when they are on heat, as well as showing the characteristic receptivity posture; what the hormone treatment has done is to make mounting more likely, and receptivity less likely. Conversely, a newly born male rat that is castrated will have the balance tipped in favour of showing more 'female' patterns in adulthood: he will tend to adopt a receptive posture, and may even be less aggressive than his more fortunate brothers who have escaped the knife. Because it is thus possible to manipulate an animal's copulatory *postures*, some people have suggested that what makes a homosexual behave as s/he does is the 'wrong' pattern of hormone secretion long before they are born.

An interesting feature of much of this literature is that this tendency to show male-like behaviour following hormonal treatment is referred to as 'lesbian' behaviour,[22] as though the two were synonymous. One of these studies not only describes behaviour more typical of one or the other sex as being equivalent to 'homosexuality', but suggests that the frequency of sexual encounters with members of the same sex is increased after hormonal manipulation.[23] At first sight, this appears to support the contention that the early hormone environment can influence later homosexual behaviour, but on closer consideration it does not. Remember that the female animals given early androgens showed an increase in mounting behaviour. Most rats, with the exception of oestrus females (i.e. females on heat), will normally repel individuals trying to mount them. So it follows that the rats that are most likely to allow mounting by this hormone-treated female will be other females who are themselves receptive. Similarly, castrated males will be more receptive than most males, and will therefore be receptive to other animals that try to mount – usually other males. The experimental manipulation, then, has made it more likely that homosexual encounters will occur, simply because it has altered the probability of occurrence of specific sexual behaviours. It tells us nothing, however, about sexual preference *per se*.

Suggestions such as these are a new variant on the idea that homosexuals have the 'wrong' pattern of hormones as adults. The implication is that, with appropriate hormonal treatment, homosexuality can be cured. These suggestions are based on a number of false premises.

First, it has not so far proved possible, by these or any other means,[24] to change an animal's – or a human being's – sexual *orientation*, despite occasional claims to the contrary. In other words, there seems to be a confusion in such research between orientation (that is, the sex of the partner that is preferred), and sexual postures more typical of one or the other sex during heterosexual intercourse.

Second, there is often an implicit assumption that human behaviour is to a large extent *determined* by prenatal or adult hormones, thus denying any contribution made by the individual's environment. At present, evidence for any direct determination of human behaviour by hormones is scanty and contradictory,[25] so it would seem rather premature to assume direct hormonal determination in the case of homosexuality.

Furthermore, the range of sexual behaviour shown by humans is enormous, and our sex drive is much less directly dependent on hormones than that of other animals. Removal of a rat's ovaries will render her asexual, but removal of ovaries in women (which often accompanies hysterectomy) has very little effect on sexuality. As a scientist who has spent some years working with female rodents, I should also mention that definitions of animal sexuality are about as limited as are concepts of human sexuality: they revolve around reproductive, heterosexual intercourse. It is a truism to state that heterosexual intercourse is necessary to survival, but need we assume that to be all? Animals often engage in other behaviours that humans include in their sexual repertoire – genital licking, or anal manipulation, for example. In animals these are called 'social contact'. In one paper using references to females of other species, it is claimed that: 'homosexual contacts between subhuman [that is, other animals] females never appear to result in orgasm'.[26] Interestingly – although the author omits to point this out – no one has determined whether female mammals other than women have 'orgasms' in heterosexual sex, either – or, if they do, what constitutes an 'orgasm'.

There have been many attempts to show that homosexuals are hormonally different from everybody else, and they have been 'mainly disappointing'.[27] To date, the evidence for hormonal differences is, to say the least, equivocal, and frequently based on shaky methods.[28] One report, for example, claimed to have found raised levels of androgens ('male' hormones) in lesbians,[29] which would, of course, accord with the mythical stereotype of lesbians as being exceptionally masculine. The number of lesbians tested was four. Despite the low numbers, as well as several other technical drawbacks to the research,[30] the report appeared in the prestigious journal, *Nature*. Another, more careful, report in *Nature* showed no physical differences (for example, menstrual cycles, body build, hormone levels, and so on) between lesbians and their heterosexual sisters.[31] This study did find some personality differences (for example, in 'extraversion' and 'neuroticism' – lesbians scoring higher on tests of these), but another study done the same year[32] reported no differences at all, either in physical or personality dimensions.

Many women are now coming out as lesbians in the context of the Women's Liberation Movement, which may present some problems for those who believe in hormonal determination of lesbian behaviour. It is more difficult to assume hormonal causes of lesbianism in a woman who has been heterosexual for some years, and then becomes a lesbian. Lesbian–feminists tend to view their lesbianism as a matter of choice. Through their involvement in feminism, they have learnt to love and value other women, and choose to express this through their sexuality. We need not worry about any possible confusion in the literature, however: if it is more obviously a woman's choice, then it can be referred to as 'pseudo-homosexuality',[33] thus making a separation between lesbian–feminists and 'other' lesbians (whose avoidance of relationships with men can then still be referred back to biology!).

WHAT DO THE EXPERTS HAVE TO SAY?

Much has been written about lesbians. Some writers are relatively sympathetic to gays and gay rights,[34] while others clearly indicate society's norms. I have selected a few quotes of the latter variety, just to give some idea of the statements made by 'experts' for other 'experts' to read:

'On the whole, homosexuals seem to be fairly ordinary people trying to cope as best they can with their particular disability. This is not to deny the psychopathological basis of homosexuality. The unconscious symbolic quality of much homosexual conduct reveals itself unmistakably in some individuals in such phenomena as compulsive assertiveness or submission during the sex act, or in either violent possessiveness or utter disregard of the sexual partner.'[35]

'One must guard against the facile assumption of many homosexual apologists that what is biologically natural is necessarily desirable or permissible. Murder does not become any more tolerable because it can be viewed as the expression of a natural impulse.'[36]

'Those lesbians who protest that, for them, this kind of relationship is better than any possible intimacy with a man do not know what they are missing. There is no doubt that, for women who, for whatever reasons, have been unable to get married, a homosexual partnership may be a happier way of life than a frustrated loneliness; but this is not to say that it can ever be fully satisfying.'[37] [It will surprise no one that the writer was a heterosexual man.] .

'. . . that to be a woman loved by a man and who has children by him is the first and most important aim of feminine existence.'[38]

'Basically all homosexuals are alike – looking for love where there can be no love, and looking for sexual satisfaction where there can be no lasting satisfaction.'[39]

'Homosexuality represents a disorder of sexual development and does not fall within the range of normal sexual behaviour.'[40]

Trainee-experts for professions that might deal with homosexuals (the so-called 'caring' professions) will read these quotations, which will reinforce any prejudices they may already have had. Some of them may eventually be in a position in which they are expected to help a homosexual person who has cracked up as a result of society's abhorrence of her/him, or who has fallen foul of the law. Many therapists are genuinely concerned to help that person overcome her/his problems, without trying to effect a fundamental change in her/him. Unfortunately, quite a large number of therapists do in fact try to change the gay person, to make her/him into a heterosexual. This does not usually work, but let us consider some of the methods used in this effort.

TREATMENT FOR THE CONDITION

Treatments are based on the notion that the homosexual is abnormal, and are therefore aimed at eliminating the 'abnormal' behaviour, and ideally instituting 'normal' heterosexual behaviour. More often than not the second aim is never realised.[41] Despite the lack of any real evidence that homosexuality is hormonally determined, treatments include hormonal as well as psychological tactics.

Ridicule or persuasion

This is used sometimes in psychoanalysis. Phyllis Chesler, in *Women and Madness*, describes lesbians being asked by heterosexual psychoanalysts: 'How could you? Women are dirty, they smell'; 'Lesbianism isn't necessary, it's absurd'.[42] If there was any confusion in a woman's mind before such therapy, it is surely likely to be increased by such statements. Lesbians sometimes end up seeking (or being referred to) psychiatric help, as a result of trying to cope with a world that defines them as deviant, that condemns their very existence. Being taught to dislike their lovers is hardly going to help.

Many behaviour therapists continue to treat lesbians as pathological (that is, sick[43]). A few are beginning to question whether it is right, or useful, to treat them as such,[44] since the therapist contributes to the social prejudice surrounding homosexuality *by behaving towards the patient as though s/he is sick*. Some patients do claim that they want to be changed from being homosexual, but this in itself may partly be a consequence of the negative attitude of the therapist, which contributes to the negative self-image of the patient.[45]

The aim of the therapy is to uncover the reasons why a person became gay, and to develop some form of 'treatment' for it. Many reasons may be suggested, some biological and some developmental; that the gay person had a 'faulty family constellation',[46] for example. One oft-quoted reason for becoming gay is one's relationship with one's mother. For gay men, for example, this means that their mothers were 'too close-binding and intimate'. But this has come in for criticism: 'What is wrong with such a mother unless you happen to find her in the background of people whose current behaviour you judge beforehand to be pathological?'[47] What, indeed? I wonder how many mothers of gay daughters or sons have been made to feel inadequate and guilty by the incredible theories of the psychotherapists?

Hormone treatments

One type of hormone treatment that has been used most on men imprisoned for homosexual offences[48] is the administration of a hormone that diminishes the sex drive. These drugs are anti-androgens, which work by counteracting the effects of a man's own testes' hormones (androgens). The hormone treatment, in fact, does nothing at all to alter his basic sexual orientation: all it does is to suppress the sex drive. Because of this, it has limited use for social control of homosexuals. Perhaps the most important ethical question it raises is the possibility of drug use being a condition for getting parole – 'you'll get out earlier if you take these tablets'. As long as the 1967 Sexual Offences Act remains, so remains the risk of gay men in prisons being thus bribed. Lesbians, of course, are less at risk from such bribery as the law does not recognise that they exist.

Another possibility for treatment has come to light recently. This is based on the effects of pre-natal hormones on the brain differentiation of animals. From such work, one researcher has proposed that the brains of homosexuals respond differently to hormones, owing to an effect of such hormones while they were still in their mothers' wombs[49] (mothers can thus be blamed again!). His biases are very apparent; he refers to the tendency for male rats castrated at birth to show a 'female' response as *homosexual behaviour* and suggests that it might be possible to 'prevent homosexuality' by administering drugs to pregnant women. He is also working on the development of the techniques necessary for such manipulation. Initially, this would involve extracting a small amount of amniotic fluid from around the foetus in order to assay its hormone levels. Then, if the hormone levels are 'too low' (if a male foetus) or 'too high' (if a female foetus), extra hormones can be given. Lest you think that these analyses are unlikely, I should add that this same author has been working on methods of determining foetal sex and hormone levels of mothers bearing male or female foetuses. The technique has some dangers and is used only if a mother is at risk of producing a genetically malformed foetus. Giving sex hormones to women in pregnancy can, furthermore, cause malformations of the foetus. However, the fact remains that work *is* going on to develop better techniques.

This scientist has tested his hypothesis that homosexuals' brains respond 'abnormally' to injected hormones.[50] Apart from the fact that his methods are highly suspect scientifically,[51] all this work is based on the assumption that hormones – whether pre-natal or adult – determine the direction of one's sex drive and affections. As I have already pointed out, there is *no* evidence to support this. Basically, his advocacy of interference with pregnancy raises questions of the rights of women in pregnancy, and, quite simply, is bad biology.

At the end of his book, he does query whether we should interfere: after all, he points out, so many artists of merit were homosexual! But he goes on to justify intervention by saying that 3 per cent of homosexuals commit suicide – ignoring the fact that the suicide rate for the whole population is not much less.[52] He also says that: '[those] with inborn sexual deviations are suffering from· psychosexual pressure'. In an age in which sex is sold to us from every magazine, every hoarding, every film, imploring us to compete and perform, there must be many people who are suffering from 'psychosexual pressure' – be they gay or not.

Aversion therapy

This is based on the assumption that it is possible to condition someone into a preference for the opposite sex, and a revulsion for the same sex. It is carried out by such methods as using drugs to make the person vomit, or to become very dehydrated; or by using electric shocks. [53] One example of the use of drug therapy entailed giving the patient (male, in this case) a drug to make him feel nauseous, while encouraging him to recount stories of his sex life with his current lover. The entire procedure was subsequently repeated, while the therapist encouraged this *fantasy* (*sic*) and used words like 'sickening' and 'nauseous' to describe the sexual acts. The object of the exercise is to train the patient to feel thoroughly nauseated at the thought of homosexual activities. Following this, the patient was given hormone injections while looking at pictures of naked women. We do not know his subsequent fate: follow-up studies of such patients are rarely done. The attitude of many of these therapists is perhaps best illustrated by a quote from a report of a treatment of a 'woman's clothes fetishist': [54] 'At one session, by a particularly *happy* chance, one of his favourite pictures fell into the vomit in the basin so that the patient had to see it every time he puked' (emphasis mine).

As usual, the literature is mainly about men; but, lest the reader imagine that it is only gay men who receive such treatments, I should add that they are carried out on women who have 'indulged' in homosexual behaviour, to encourage them to behave completely heterosexually. [55]

Electric shock therapy has also been used, either through some part of the body, including the genitals, or through the brain (although the latter is less used now). [56] These treatments rarely achieve the intended result of knocking out the homosexuality and implementing heterosexuality, and have been known to do the reverse.

What kind of caring therapy is it that makes a human being vomit, or feel fear (associated with previous electric shock) at the sight or thought of a former lover? Jane Rule, in her book, *Lesbian Images*, [57] also points out that no consideration is ever given to the former lover, who has to face the world deprived of her love, and having to watch the vomit-strewn suffering of the victim who once loved her.

These treatments, most of which are derived from the results of animal experiments, are punishing. They are used to control and to punish the gay person for daring to flout society's conventions. Gays are feared, which gives rise to loathing, so they must be kept from the public eye and forced into a ghetto existence: [58] the 'I-don't-mind-gays-but-wish-they-weren't-so-blatant' [59] attitude prevails.

Punishment for homosexuality is nothing new, as I have already pointed out. Once, it was a heresy, requiring burning at the stake. In Nazi Germany, lesbians and gay men were singled out, and their houses marked with a pink triangle: most were sent to the gas chambers. Gay people, and especially lesbians, are continually harassed, denied or thrown out of jobs, lose custody of their children, are in constant danger of attacks. The punishments are everywhere, but it is in the context of 'treatment' that they are raised to the status of 'scientific' procedures.

Both the use of scientific statements to give credibility to the prejudice that lesbians and gay men are 'abnormal', and the use of punishments, serve to maintain a system of oppression, in which gays are oppressed specifically because of their choice of sexual partner. Lesbians are doubly oppressed – both as gay and as *women*. In patriarchal societies, men are taught to expect to have access to women's bodies, and to control them:[60] by their very existence, lesbians are denying that control, so they are punished.[61]

The advent of the Women's Liberation Movement and Gay Liberation in the 1970s has made many more people unafraid to admit that they are gay, and has created an environment in which women can have more choice about whom they wish to spend their time with. To some extent, these social changes create possibilities for changing prejudice, but we have to go beyond simply removing prejudice. It is not enough. Scientific arguments are being used increasingly to justify a whole range of power relations in our society – men against women, whites against blacks, straights against gays. The straight, white, middle-class male is the standard against which all others are judged: and found wanting. The scientific arguments used to 'prove' that lesbians are abnormal, maladjusted, hormonally deficient or whatever are just another example of this misuse of scientific method to maintain the *status quo*. It is time we changed it.

Lesbians *are* different from heterosexual women: they have chosen not to relate to men. For lesbians in the Women's Liberation Movement, that is less a sexual choice than a political choice. It is precisely this that many people find threatening about lesbianism within feminism. That some women wish to extend their love of other women to include sexuality should be a matter of their choice, made without fear of reprisal. Unfortunately, there is considerable risk of reprisal. Those who would condemn homosexuality usually base their prejudice on ideas that homosexuals are suffering from proven abnormalities, a prejudice that the scientific literature continues to feed. Scientists will no doubt continue to attempt to find an alternative answer, as the old ones fail – alternative answers which may provide new ideas for trying to keep lesbians under control.

Or we could fight back.[62]

'Sickness is a Woman's Business?':
Reflections on the Attribution of Illness

Hilary Standing

Many people must be familiar with the statistic that one in six women, as opposed to only one in nine men in Britain, may expect to spend some time in a psychiatric hospital. The frequency of consulting GPs is also higher among women, whether for mental or for physical conditions.[1] In a gender-conscious society, it should be unnecessary to point out that such statements provide a frame for value judgements about women and may strengthen stereotypes about the 'weaker sex' or the 'hysterical nature' of women.

The aims of this chapter are, first, to look critically at the basis for the statement that women are more ill, or present themselves as more ill, than men; second, to consider arguments used to explain the difference by comparing them with material from other cultures; third, to demonstrate, by considering these explanations, some of the ways in which debates about the mental and bodily states of women can be turned into controlling devices operating, as the psychoanalyst Karen Horney points out, 'to reconcile women to their subordinate role by presenting it as an unalterable one'.[2]

The material upon which I draw relates mainly to our category of 'mental illness', but it should be borne in mind that the dividing line between mental and physical illnesses is often an arbitrary one and that a condition such as depression is quite likely to be expressed through or result from physical symptoms. The tendency to see aspects of 'female' behaviour as the inevitable outcome of hormonal changes is a further case in point. It is, perhaps, less often realised that *physical* illness rates also show much variation by sex, social class and occupation.

Some problems of definition need first to be discussed. I approach the issue as a social anthropologist interested primarily in the meanings and values that different societies and social groups attach to illness states. I am not concerned or competent to judge whether or not the findings of biologists and medical scientists are 'correct'. The concern of the social anthropologist is with the social messages that states of illness may carry. An illness such as 'depression' carries a much greater load of these messages and judgements than does, say, tonsillitis or malaria. It follows, then, that to attribute more mental illness to women than to men carries with it greater opportunities for making judgements, often of a stigmatising kind, about the 'nature of women'.

In order to approach this social dimension of illness, two distinctions are necessary. First, we should distinguish between sex and gender. Ann Oakley has said that 'gender is a matter of culture, it refers to the social classification into masculine and feminine'.[3] Gender is the social aspect of sex. The distinction between gender and sex can be demonstrated by taking, for example, hospital admission statistics which are broken down by sex. This is a biological classification which does not carry within itself any statement about the nature of maleness and femaleness. Compare this with the statement, 'women are more susceptible to stress illnesses than men'. This is a statement about *gender*, in that it makes an assumption about the different *social* constitutions of males and females. Explanations of different illness rates tend to be statements about gender, or in any case about the experience of femaleness.

The second distinction is between disease and illness. It is quite possible to have a disease without feeling ill. Disease refers to a pathology. Illness is a broader term in meaning which may cover (and for my purpose, will cover) the recognition of sickness and the behaviour associated with it – or what sociologists have called the 'sick role'. Illness, for people the world over, is primarily a social happening. It affects relationships, disrupts production and may drastically affect the standard of living of ill people and their dependants. 'Being ill' can thus depend on practicalities: 'Can I afford to take time off work? Can I leave my husband to prepare the meals?' and so on. Illness statistics, though appearing in the language of 'hard data', are dependent on such subjective judgements.

The distinction between disease and illness is, in any case, a feature of the Western medical view of sickness. The medical systems of other cultures tend to take a more 'social' view of the nature of illness. This makes comparison across cultures a difficult matter, but

I would argue that, in the context of illness, the similarities are at least as great as the differences, in the sense that the life experiences of women who suffer from depression, for instance, share an underlying pattern, whether they are depressed in East London, South India or Taiwan.

This leads to a further difficulty that arises whenever a medical model is used to talk about states such as depression, or when we use statistics on drug intake to illustrate the 'sickness' of a population. In 1972–3, more than 30 million prescriptions were dispensed for tranquillisers, antidepressants, stimulants and appetite suppressants.[4] Boredom, poverty, bad housing and so on have become medicalised. The handing out of pills is frequently a substitute for considering treating the primary causes. But it could also be argued that there has been a real increase in the amount of stress experienced in the population, particularly by women, who account for the greater number of prescriptions dispensed.

We may now turn to the statistical differences in illness rates for males and females in the UK, bearing in mind the conclusion of three leading epidemiologists[5] that, 'in spite of gross variation in the definitions of illness and incapacity, and the different measures of frequency used, rates for females are consistently and substantially higher than those for males'.[6]

The main sources of published information are statistics on hospital admission and length-of-stay rates, which are produced annually by the Department of Health and Social Services. These are broken down by sex, age, diagnosis and frequency of admission. They tell us not so much about the extent of illness in the population, but about who gets referred to the psychiatric services. A different kind of source, which tells us about people's own judgements as to their state of health, is provided in the General Household Survey, which covers annually a small random sample of households. This contains a section on doctor consultation, self-medication and time taken off work in a two-week reference period. Unlike hospital statistics, this information is broken down by social class and by marital status.

These statistics, like most, have to be treated with great caution. They can pose major problems of comparison. Sometimes figures for England and Wales are aggregated. At other times they appear separately. Figures for total hospital admissions may be broken down into first and subsequent admissions, or they may appear as an aggregate. Diagnostic categories change from time to time so that figures for different time periods may give a misleading picture of changing trends. Admissions policies also change. New forms of care have resulted in a decreasing hospital population.

Although the crude rates show a clear difference between the sexes, they need to be broken down in terms of other factors in order to reach less crude conclusions. Age, social class and marital status all affect the rates. For instance, the total admissions rate shows the highest peak for either sex and all age groups for women in the 'middle-age' category 45–55. If we look at the self-reportage rates in the General Household Survey, however, we find that single women in that age group report the lowest rates of any marital status group of either sex. If we compare single men and single women, we find that single men aged 35 upwards have higher rates of psychiatric admission than single women, rising to nearly double the number in the age group 65–74. The more sensible units to compare here are age and marital status, rather than sex alone.

Furthermore, if we take the age factor into account we get a more complex picture. Up to the age of 15 admission rates are higher for males, but total admissions for all young people are much lower than for older groups. From 15 upwards the rates increase sharply and show their peak for males in the age group 25–35 and for females in the group 45–55. For males, the rates then decrease gradually with increasing age, and for females they also decrease between the ages of 55 and 65, but then increase again to the extent that there are more than twice the number of women aged over 65 in psychiatric beds than there are men.

A few points might be made about the effects of age on the general population structure, and therefore on the number of hospital admissions. About one-fifth of psychiatric admissions are of people aged 65 and over. The life expectancy of women in the UK at birth is presently six years longer than that of men. This means that there are somewhere in the order of 12 million more women in the total population. Thus, women outnumber men, particularly in the over-65 age group. As there is still a tendency for women to marry men older than themselves, we may account for some, at least, of the substantial difference in admission rates by looking to population and social factors. The chances of senile men being looked after at home by younger and fitter wives are statistically much greater than they are for senile women to be looked after by husbands. The tendency for elderly women to swell the hospital statistics shows up very clearly in the figures for the numbers actually resident in psychiatric beds. In the over-65 age group there are three-and-a-half times as many women as there are men. Increased life expectancy, although favouring women, does not favour them socially or with better health.

Diagnostic variations present other problems in assessing statistics. Diagnoses are subject to changing medical fashions, particularly in the field of mental illness. Statements such as, 'schizophrenia afflicts women more than men' should really read, 'at various times, more women than men have been diagnosed as schizophrenic'. In the last five years, the proportion of diagnosed schizophrenics has fallen for both sexes, but more so for women, and there is now a negligible difference in the rates for this diagnosis. The greater numbers of women resident in hospital as diagnosed schizophrenics reflects the greater fashion for this diagnosis in previous years and the tendency for these patients to remain in hospital for much longer periods of time.

The category of depression [7] must be treated with a similar caution, especially as higher rates are generally associated with women. In 1970 diagnostic categories were changed in accordance with the Eighth Revision of the International Classification of Diseases. Prior to this, diagnoses of 'depression' were put in the category of 'depressive psychoses' – a category in which women outnumbered men by two to one. Since 1970, in accordance presumably with a desire to distinguish between different types of depression, diagnoses that say simply 'depression' have been transferred to the category 'all other conditions'. This has resulted in the reclassification, since 1970, of no less than 35,000 psychiatric cases!

Despite these difficulties, are there nevertheless grounds for accepting that women are more vulnerable to mental illness than men? The age factor has already been mentioned as one qualification to this view. Another likely explanation for the higher overall use made of the health services by women lies in their reproductive role. Data for the UK do not distinguish medical conditions associated with reproduction, but a study of the North American rates suggested that this made a difference of 20 per cent to the overall rates – enough to make a significant dent, but not to cancel out the difference completely. Unfortunately, it is not really possible to assess very accurately the effect of reproduction on rates of mental illness. Cases of postpartum depression (that is, clinical depression within the first few days of childbirth) are not given separately, and it is even more difficult to evaluate the indirect but crucial effects of childbirth and childrearing on the mental health of women. That, however, they play some part in triggering depression is apparent from a number of studies (see below).

A further suggestion that would perhaps make a dent in the

difference between the two rates is that of the 'masking' effect. The idea here is that men and women are equally vulnerable but that women turn to antidepressants whereas men turn to alcohol. Women need to get prescriptions from their doctors, whereas men can obtain their alcohol over the counter, and hence do not appear in the statistics on drug intake unless admitted to hospital as alcoholics. This is not an easy hypothesis to test. It involves judgements about normal and abnormal alcohol intake, which are just as much social judgements as medical ones. If hospital admissions for alcoholism are taken into account, it should be noted that these account for a small proportion of the total intake, and that the rates for female alcoholism are increasing proportionately slightly more than they are for men.

It is impossible to draw any firm conclusions as to whether women do or do not have more illness than men from taking the statistics as they stand. I have mentioned a number of areas where the figures should not be taken at face value, but there is a still more fundamental problem. To answer this question would require us to make a distinction between a 'real' disease and the experiences that people have, which they call illness. In everyday life it is the latter that is crucial. To report oneself ill is to make a significant social statement, and what the figures do suggest is that we should look particularly closely at those age groups and statuses among women that show the greatest rates. It is in this context that material from other societies can be most useful for reference.

A number of societies that have been studied by anthropologists explicitly associate illness with women. Harriet Ngubane, discussing Zulu medicine, reports that Zulus say, 'illness is mainly a woman's business'.[8] The pagan Hausa of northern Nigeria maintain that 'only women become ill'.[9] But in Western medical terms, Hausa men suffer more serious illnesses from which they are more likely to die, and at younger ages, than do women – an observation that seems to parallel the Western case. It is necessary to look more closely at the category 'women'. In both these cases, 'women' encompasses 'women and children'. For Zulu, illness is a woman's business because so much of it concerns unborn babies and young children. Children both suffer more actual illness and are considered more vulnerable than adults to a whole range of conditions.

The lumping together of women and children is even more evident in the Hausa case, where illness is that which happens in the home, the domestic sphere of women in which men are not particularly concerned. In our own society the picture is not dissimilar. Illness is very much a mother's business – children are far more likely to end up in a doctor's surgery with a female parent than a male. If illness depends, to an extent, on self-reporting, women do at least have more opportunity to do so. Women are also more vulnerable to illnesses contracted from children. Perhaps more significantly, women are more involved in a host of preventive screening measures to do with their reproductive role or the health of their children. The Hausa mainly define illness in practical terms. It is that which stops you from working. Similarly, a North American study found that working mothers tend to report less illness than mothers who remain at home. Illness, for ourselves no less than the Hausa, means loss of production. It is that which requires a worker to obtain a medical certificate to stop work.

I now turn to consider some of the explanations that scientists and social scientists have put forward to account for the assumed difference in illness patterns between the sexes. It might be said that the 'correctness' of the statistics is irrelevant here. Some of the explanations have been formulated explicitly to explain the statistical difference – in other words, their authors accept that the statistics give a correct picture of the situation. Others have been formulated out of separate research which has confirmed the difference. They are therefore of interest both as explanations that can be tested and in their views of 'femaleness'. The explanations that I consider here are not necessarily exhaustive, nor are they necessarily incompatible. Some are clearly contradictory, however. I list them all here and then deal with each in turn.

1 Women suffer from greater vulnerability to hormonal stress which results in more incidence of illness, but does not affect mortality rates.
2 Women have more illness than men because their assigned social roles are more stressful.
3 The greater amount of illness among women is directly related to what has been called 'severe life events' such as sudden unemployment, eviction, bereavement· and so on, all of which, it is suggested, take a heavier toll in stress among women.
4 Women report more illness than men because it is socially more acceptable for them to do so.
5 The sick role is more compatible with women's other role responsibilities.
6 High illness rates, particularly for depressive conditions, reflect women's over-socialisation and over-playing of the feminine role.

The first explanation – that women are more vulnerable to illness through hormonal stress – is difficult to evaluate. Biologists are themselves in disagreement about this. One of the most well-known arguments proposing this link is Katherina Dalton's study of the menstrual cycle.[10] She holds that women are statistically more likely to take time off work, have accidents, be admitted to psychiatric hospitals and commit crimes or make suicide attempts in the ten days preceding menstruation. Her findings have, however, been criticised on the grounds of statistics.[11] Arguments about the link between hormonal changes and behaviour are far from being settled. Such arguments have great potential for being turned against women. Unwelcome behaviour may be dubbed 'irrational' and blamed on 'the time of the month' and so on. Looking at non-Western and non-industrial societies, there is little evidence of the existence of a pre-menstrual syndrome (see, for example, Chapter 5). This does not, of course, mean that the syndrome does not exist elsewhere, only that if it does exist it has either gone unreported or the symptoms reported are different.[12] It should be pointed out, however, that in societies with high fertility rates and long weaning periods, women experience far fewer menstrual cycles than women in the West.

The other five explanations all make the assumption that some at least of the illness afflicting women is related to gender roles. On the face of it they seem to be saying rather similar things. In fact each one uses the concept of role in a quite different way.

The second explanation claims that women have more illness than men because the roles assigned to them as women are more stressful than those assigned to men. 'Role', in this context, means the role created by marriage, which, it is suggested, makes greater and more diffuse demands upon women. Men, on the other hand, appear to experience a degree of protection from illness by marriage, the rates for married men being lower than for single men and for women, both married and unmarried. The rate for married women is the highest of all the categories. This argument is suggestive, but not particularly well developed. It assumes, rather than demonstrates, its claims. It would be difficult to test in a comparative way because in many societies women have little choice about marriage: it is an inevitable aspect of maturity and the meaning and obligations of marriage vary between and within societies.

The third explanation claims that there is a causal link between psychiatric disorder and what the author calls 'severe life events' or life crises affecting women. This is contained in a set of studies by George Brown and Tirril Harris on class, gender and psychiatric disturbance in a district of south London.[13] In these studies, depression is treated less as a clinically defined medical condition than as a fact of life for many women. The study first established the frequency of depressive conditions in the district by using standard psychiatric measures for depression,[14] irrespective of whether the individuals concerned had been formally labelled as psychiatric cases or not. The aim was to establish a 'truer' rate of incidence than is to be found in looking only at the cases already referred for treatment. A comparison was made between working-class and middle-class women and between women undergoing treatment for depression and women who were judged by the researchers to be depressed but were not undergoing treatment. The findings were that five times as many working-class women as middle-class women were chronically depressed and that this depression could be correlated with severe difficulties experienced by working-class women in such matters as housing and unemployment. The burden of these problems was seen to fall more heavily upon women.

To give a very brief summary, the study showed the highest risk category to be the young working-class married woman with young children at home and no outside employment. It might be summarised even more briefly by saying that deprivation and depression are clearly linked. The study also found that young women in all social classes are the most vulnerable to the kinds of life crises that may trigger depressions, a finding that fits with the psychiatric consultation rates and which suggests that it is appropriate to consider not only women's roles but their life situation and relative disadvantages as well.

Psychiatric consultation rates are highest in the West for women in the 25–35 age group, and in the group 45–55. The most clear 'severe life events' affecting the younger age group seem to be poverty and childrearing. To this we might add the peak in the divorce rate which occurs between the ages of 25 and 30. Divorce, particularly for the dependent woman with young children, often affects women more adversely than men.

The most common argument for the higher rates in middle-aged women is the menopausal depression syndrome. This argument is as much an argument about the symbolic significance of the menopause as about the hormonal and biological changes. In a society that places disproportionate value on a woman's youth, and hence her ability to 'attract' men, and then defines her role primarily in childbearing and childrearing terms, the menopause signifies the end of this reproductive phase and tends to coincide with the departure of grown children.

To what extent are women 'victims' of life crisis situations such as these in other societies? There is a lot of anthropological evidence to suggest that there are similarities in the pattern of women's lives in these respects. A striking comparison with Brown's study of working-class London women is to be found in a study by Byron Good of concepts of illness in an Iranian town.[15] He found that one of the most commonly experienced and reported illnesses, particularly among poorer people, was 'heart distress'. This was not, however, the same thing as our 'coronary'. To the sufferers it was explicitly a response to worries and problems. In our own cultural context, we would probably describe it as 'depression'. The symptoms experienced are different but the conditions are remarkably similar. Good reports that the incidence of 'heart distress' is highest among women of lower social class and of childbearing age. Seventy-three per cent of his case sample were in this category. The evidence is poignantly summed up in this quote from a case study of one of the women: 'We are poor, we don't have money, we all have heart problems.'

Turning to a different context, a study by Mark Nichter of high-caste Brahman women in south India[16] highlights the very considerable stresses experienced by these women who are treated all their lives as dependants of men, bearers of the all-important sons, and guardians of the status of their caste through the elaborate food preparations and purity rituals that they are required to carry out. A study of the local traditional medical practitioner showed that his clients were largely drawn from the high castes and consisted of many more women than men. These women fell mainly into two categories – young women of marriageable and childbearing age, and older, usually widowed, women. The majority of complaints fell into what we would describe as psychosomatic illness – weakness, irritability, burning sensations.[17] Menstrual complaints accounted for two-thirds of the symptoms reported. Nichter suggests that these high-caste women experience a high level of stress resulting from the conflicting demands upon young women to behave in accordance with their traditional roles of wife to a man chosen for them by their parents, but at the same time also to take advantage of the educational opportunities now open to them. Thus education becomes not a means to independence, but a passport to a 'good' marriage. The illnesses experienced by these women may also be seen as one of the few responses available to them in an oppressive environment.

Turning to the other major peak in women's lives, that of the menopausal years, one finds rather less evidence for its 'crisis' aspect in anthropological accounts. In fact, very little attention is paid to this stage of the life cycle as compared, say, with the beginning of menstruation and with childbirth. One possible reason for this may be found in the figures on life expectancy. Western women may expect to live up to one-third of their lives after the menopause. This is not the case for women in the Third World, where the age structure of the population is disproportionately young. However, elderly women do exist in all populations, and with mortality rates generally favouring women (India is a notable exception) there are usually more elderly women left alive than there are men. Another suggestion made by anthropologists to account for the apparent lack of attention to the menopause in many societies is that menopause frees women from the burdens imposed upon them by the menstrual and childbirth taboos and allows them to play a greater rather than a lesser role in activities normally monopolised by men. On the other hand, it should be pointed out that the lot of elderly and widowed women is often a very unhappy one.

One study that does note a clear link between menopausal women and the onset of depression is the set of Ghanaian case histories given by Margaret Field, a Western-trained psychiatrist practising in rural Ghana.[18] The case studies are classified according to standard psychiatric categories. Out of her twenty-seven cases which she classifies as depressive, twenty-six are women of around menopausal age and described by Field as 'conscientious women of good personality who have worked hard and launched a fleet of well brought-up children'.[19] In addition, 'many of them have paid for their children's schooling with money earned by diligent trading, market gardening or cocoa farming.' Field sees a major reason for the incidence of depression among these women to lie in the social hazards faced by this age group. Husbands may take a younger wife upon whom more attention is lavished; attention not only taken away from the older woman, but also often paid for out of the older wife's labour. The younger wife is fertile. She can bear yet more children to the ageing husband. For the older wife, her reproductive life is over – a dilemma summed up in the menopausal woman's complaint: 'I am pregnant, but the pregnancy doesn't grow.' The readjustments that have to be made by women at this age contrast unfavourably with those made by men at the same age.

This study makes an interesting comparison with Western studies documenting menopausal depression. It might be suggested that in societies that place great emphasis on fertility, or in which a woman's primary role is in childbearing, menopause will be a time of particularly severe crises. Pauline Bart's study of depression among middle-aged American women illustrates this rather clearly.[20] Her sample was mainly of women who had devoted their lives to child-rearing and who felt that their reason for being alive had gone when their children left home and they were no longer able to bear any more. It could be argued that the United States, or possibly Western capitalist society generally, represents an extreme in apportioning child care to women, and that in other societies women do also play important productive roles that do not cease at the menopause, and at least make room for a less narrow definition of female sexual identity than that imposed on the conforming American woman.

Other factors making life difficult for women at this particular stage include bereavement. The death of a spouse is a more common experience for women than it is for men. In the UK, for instance, women lose annually 96,000 more spouses per year than men. It is in the age group from 50–60 that the numbers rise sharply. Nichter pointed out that Brahman widows formed a significant proportion of the traditional healer's clientele. This was connected, by both patients and practitioner, with the feelings of loss of value and neglect following the loss of both husband and independent children. It should also be said that in high-caste Indian society, as in many others, a woman's identity is completely submerged in that of her husband. She has few rights of her own and her existence is defined in relation to men – she is someone's husband or daughter or mother.

Finally, a study that illustrates well the peaks in female life crises is Margery Wolf's study of rural Taiwanese women.[21] She looked at suicide statistics for women over a period of years and found that they 'dramatically reflect the crisis periods in a woman's life'.[22] Women in their early and middle twenties produce the highest rate for any group, a fact that she attributes to cruel mothers-in-law, but which is more likely a response to a whole range of anxieties involved in getting used to marriage, and a fear of infertility. The suicide rate then drops sharply, only to rise again for women in their forties and fifties, reflecting, says Wolf, 'the crises created by marriages that intrude strange women into the beds and affections of sons'.[23] Ageing women become aware, once again, that their 'usefulness' is dictated by their capacity to reproduce.

Three more explanations remain to be discussed. These might all be termed, in different ways, strategy explanations: (1) women can more easily 'play the sick role' because it is more compatible with their other role responsibilities; (2) higher illness rates reflect women's 'over-socialisation' into the feminine role; (3) women report more sickness because it is socially more acceptable for them to do so. Again we have several quite different uses of the term 'role' here used. The first of the three suggests that women, or more specifically housewives, take on a sick role rather as they take up knitting – it fits in more easily with their other responsibilities, and they do not need a medical certificate in order to play this role. But to sustain such an argument it also has to be argued that women's roles in the home are less demanding than those of men or of women outside the home. Such an argument thus contradicts the claim made in a previous explanation, that women's roles in the home are more demanding, as well as an argument put forward elsewhere by the sociologist Talcott Parsons, that the illness of a woman is more disturbing to her family than is the illness of her husband. It must also be pointed out that this kind of strategy explanation sustains a view of women as 'neurotic' and prone to trivial complaint. The use of the expression 'playing the sick role', although supposedly a neutral sociologists' term, suggests a view of illness as a game indulged in particularly by women.

To take next the argument about the greater social acceptability of illness for women, this is based on findings that psychiatrists, far from being neutral in their evaluation of symptoms, are in fact influenced by gender stereotypes as to what is normal and abnormal behaviour. 'Aggressiveness', for instance, may be an acceptable characteristic for a male client, but defined as abnormal and needing treatment in a woman. The argument that has, again, been advanced by an American sociologist goes on to suggest that illness is more stigmatising to men because they are defined as the stronger sex. As the 'weaker' sex, illness is more acceptable (normal?) for women. This argument can, and should, be turned on its head. By suggesting that illness is more 'acceptable' for women it brings us back to a neurotic, scheming, female stereotype. The evidence that psychiatrists evaluate male and female patients in accordance with ideologies of what is acceptable as male and female behaviour suggests, instead, that acceptable female behaviour is defined much more narrowly. Women can get away with less variation before being labelled 'abnormal' by the standards applied to them, and these standards have nothing to do with medical categories. They are disguised ideologies about how women ought to behave.

The third of this set of arguments suggests that higher psychiatric illness rates reflect women's 'over-socialisation' into the feminine role. Here, the use of the term 'role' suggests a particular ideology about gender and the expectations that it is acceptable for women to have in their adult lives. Pauline Bart's study of middle-aged depressives is a good illustration.[24] She argues that those women who have accepted most completely the definition of women's role as dependent married partner and bearer of children are most at risk from menopausal depression because they are placed in a double-bind. Social and economic pressures combine to enforce such conformity upon women, yet the culture devalues the conformity. Women are described as 'only housewives'; as unproductive in a society in which production and competition are valued above all else.

One theme that runs throughout these various explanations is the great concern with women's bodily states. In much anthropological literature, women appear as little more than a collection of bodily processes of menstruation, childbirth, and so on. This concern may be seen in two different ways. It may be argued that this reflects the bias of predominantly male researchers who concerned themselves entirely with men and with male views of society, and neglected to discover whether women had autonomous or alternative views. On the other hand, it could be argued that, given women's exclusion from many dominant political, economic and social concerns, they have fewer ways of expressing their identity distinct from one defined by bodily processes. In other words, while men may view women as the 'weaker sex' and as 'sickly', women may be evaluating their own bodily states in positive ways.

To give just two examples: among the Hausa, Murray Last[25] found that women would declare their own menopause and that such a declaration is connected explicitly with controlling male access to women, in both the sexual and social sense. Menopause may be a welcome means by which women define and separate themselves off from men.

Our knowledge that women may well perceive their bodily processes differently than does the medical profession is illustrated in a study of young American women's attitudes to menstruation. It concludes that 'the experience of menstruation as reported in the literature has over-emphasised the negative and debilitating aspects, and has ignored the positive aspects'.[26] It would, perhaps, be extraordinary indeed were a phenomenon unique to the female sex not given some positive value by women themselves.

As with most apparently enlightening statements, however, this view contains the seeds of another kind of oppression. The Women's Liberation Movement has done a great deal to free women of the mystification, fear and taboo surrounding their bodies. We are now witnessing an increasing emphasis on women 'getting in tune with' their bodies, being positive about their menstrual cycles, and so on. Some of the recent literature on this last topic has been pointing to anthropological literature to illustrate the claim that 'primitive' (that is, pre-patriarchal) woman was treated as 'sacred', as being 'in touch with nature' and therefore powerful in relation to men. [27] But there is no evidence at all for these assertions. The sacred and the polluting are two sides of a coin. They signify social control as well as power. The exhortation to be positive about our bodily cycles is in danger of producing new mystifications based on myths about women's psychic communion with the moon or with nature. [28]

7

The Obsessive Orgasm:
Science, Sex and Female Sexuality

Wendy Faulkner

Sex appears to be the topic of our age: it represents the animal in us; it's what makes the world go round; it is naughty, perverted, dirty, exalted; it is the crux of our personal happiness; it is better not thought about.

All these things are said about sex today. All reflect a mass of values and associations that often appear to contradict one another. Indeed, the shroud of silence that is gradually slipping away from the subject has in many ways left us none the clearer as to what it is that sex is all about.

WHAT SCIENCE HAS TO SAY ABOUT SEX

Science is a triumph over ignorance and superstition – or so it goes. The picture is not so rosy, however, when it comes to understanding how women 'tick' sexually. Sex research historically has generally assumed a diagnostic and even moralistic stance. People had long known that sex eventually led to procreation, and the church decreed that it ought to. However, even the more independent thinkers of the Renaissance could not clarify the exact relationship between the two: they realised that women's periods stopped during pregnancy, but the role of the male in conception and the origin of the female egg remained a mystery.[1] Their successors in the nineteenth century finally solved the problem, but these same doctors also practised all sorts of barbarous acts on those women who had the courage to express their own sexuality. I shall be returning to this sorry saga later.

It was, in fact, only in the 1930s that sex, and sexual responses as such, began to be studied seriously. Initially, the National Committee for Research into Problems of Sex was set up in the United States, its implicit purpose being to 'solve' *social* 'problems' like homosexuality. It took another ten years, until after the Committee had freed itself from financial ties with the Rockefeller Foundation, for the more independent work of Kinsey and others to become possible.[2] The historic work of Masters and Johnson, which was started in 1954 also in the United States, was really the first research launched into the actual physiological changes and stimulation involved in sexual arousal.[3]

The research falls loosely into three categories: studies of animal sexual behaviour; direct physiological measurements of human sexual experiences; and cross-cultural survey studies of human sexual behaviour in different societies. It reflects a history of prejudice. It is characterised by an imperative to explain sex in terms of reproduction, by an ever-present confusion of the social and the biological and, above all, by a limiting, impersonal and male-orientated definition of female sexuality.

I intend to cover the major scientific work in these three fields and to show that, yet again, important questions are not asked and that often the resulting interpretations are blinkered. In posing some new questions or venturing some new interpretations, I do not pretend to have all the answers or to speak for all women everywhere. I comment simply as a woman who, like many others, is trying to identify her *real* sexual feelings from the morass of confusion and expectations.

THE FEMALE ANIMAL

Sexual reproduction in most animals requires two sexes. The genes from both a male and a female of the species are necessary for the creation of a new individual. In evolutionary terms, therefore, the sexes are considered incomplete, and sexual behaviour may be narrowly defined as that which will increase the likelihood of copulation and hence fertilisation. I want to consider sexual behaviour in female animals first, because many of the assumptions and interpretations that are made have consistently been carried over into work on women's sexuality.

Virtually all female sexual behaviour is considered to be a *response* to the male who is doing the impregnating. 'Randiness', as we may know it, is given a string of rather euphemistic titles like 'receptivity' or 'acceptance'. In other words, the female is seen almost solely as the receptacle of male sperm. The only quantitative measure used in most studies of non-primate female animals is the 'lordosis quotient'. Lordosis in mammals, the 'fully receptive position', is when the female stands quite still, bottom in the air and tail to the side. The details of copulation itself are described at length, from the point of view of the male: penetration, the number and frequency of thrusting, position and the so-called refractory period have all been measured in the male captive. The lordosis *quotient* is simply the number of times she assumes the position divided by the number of times the male successfully mounts her. In other words, it is a comparison of her passivity with her activity: 0:0 means that she is totally unreceptive and 0:8 is the average maximum of her receptivity, normally coinciding with the rise of hormones in her bloodstream at ovulation.[4] One might be forgiven for wondering if the female felt anything at all before sticking her bottom proudly in the air at the appropriate moment!

The assumption that the female animal is aroused only as a result of her hormones and directly in response to the male appears narrow to say the least. It is known through experimentation that lordosis can be induced in many animals by stroking their sides and genitals manually, and yet it is assumed that the response is always directed towards heterosexual coitus and pregnancy. In higher mammals, specifically apes and monkeys, female sexual behaviour has in fact been shown to play a social as much as a reproductive role. The 'presenting response', in which the female stands in front of the male almost offering her bottom, is not always a prelude to coitus or a sign of 'natural female receptivity'. It can be used as a signal of friendship or subordination by both females and males in the social group.[5]

In coitus itself most female animals are bound to be passive. It was not, however, until the 1970s that the idea of females *actively* 'leading' the male on to coitus was even entertained. At oestrus, the female is said to become 'sexually stimulating *to the male* and willing to mate with him'.[6] Oestrus coincides with a peak of general activity supposedly geared to facilitate copulation, and yet the possibility that some, at least, of this extra energy is used up by a sexually excited and inviting female before copulation even takes place is not considered. The licking of the female's genitals by the male dog prior to copulation is interpreted solely as a sign of male attraction to the female odours present at oestrus. Is it really so unlikely that this very action *also* excites the female dog? And if not, why is this not considered important to our scientists?[7]

Some scientists, notably R. L. Doty and Frank Beach, have shown a bit more respect for the integrity of our animal sisters. Accepting the fact that female primates enter into coitus more often at ovulation, Ford and Beach caution their readers not to make what amount to male-orientated assumptions: 'Coital behaviour in apes does not reflect exclusively the desire of the female for copulation. It simply shows when coitus took place, and we have seen that in some circumstances the males may control this event.'[8]

In a more recent paper, Beach has actually gone some way to describing sexual desire and eroticism in female animals, granting them the potential for independent and initiating behaviour. He distinguishes three categories of behaviour: sexual 'attractivity', 'receptivity' and what he calls 'proceptivity'. 'Attractivity' describes changes in the female that attract the male. 'Receptivity' is (more usefully) defined as behaviour 'necessary and sufficient for male success' in copulation. This much is not new. What is, is his concept of 'proceptive', or active female, behaviour. Beach differentiates sex in which the female is passive (that is, where she is largely responding to various stimuli), from sex in which she either stimulates or actively increases the probability of male sexual arousal. (The distinction has in fact now been identified by scientists working on sheep, monkeys and apes.) He points out that this behaviour is probably more learned and therefore less tied to hormones – like our own. For example, active behaviour in rats decreases when they copulate in unfamiliar surroundings, though they remain 'receptive'. The same is true when female dogs copulate with an 'unfavoured' mate.[9]

I would still dispute that all of these types of behaviour should always be described in reference to the male. It implies that female animals do not have a sex drive independent of males. This is a nonsense if we remember the recorded incidence of female masturbation in both non-primates and primates. Some female animals go to enormous lengths to find objects that will excite them sexually, and primates specifically use much the same methods as women. Yet despite this, our brother scientists seem to feel totally justified in concluding that masturbation never produces the same 'climactic results' for the female as it does for the male. They even go as far as to imply that the reason why, for example, some females drag their vulva over sticks and stones on the ground is to leave a trail of smell which – you have guessed it – is stimulating to the male.[10] How can they tell that females don't masturbate for their own sakes?

Beach, nevertheless, leaves us a telling indictment of the gap in most sex research:

> Proceptive behaviour is functionally as important as other patterns traditionally termed 'receptive'; but the female's tendency to display appetitive responses finds little opportunity for expression in laboratory experiments which focus exclusively upon her receptive behaviour or upon the male's execution of his coital pattern. The resulting concept of essentially passive females receiving sexually aggressive males seriously misrepresents the normal mating sequence and encourages a biassed concept of feminine sexuality.[11]

This is a bias permeating all scientific thinking on female sexuality and, as I intend to show, one that is reflected in research on women.

FEMALE SEXUALITY

Scientists have generally been at a loss to appreciate or understand female sexuality. Presumably, women represent just another object of study to that predominantly male world. It nevertheless strikes me as extremely funny to read comments like the following in a textbook coverage of post-orgasmic changes in women as compared with men:

> [In women] this apparent lack of gross change and the difficulties associated with obtaining adequate reports from females on the oc-currence of orgasm, has in the past prompted much speculation but little specific and reliable information about the nature, or even the existence, of the phenomenon.[12]

The ignorance, it seems, is total! Even Ford and Beach cautiously comment, 'The human female *appears* capable of marked sexual excitability without the physiological stimulation provided by the . . . hormones.'[13] The same statement applied to men, while equally true, would I am sure be an unqualified assertion! My point is made. Ignorance has prevailed for centuries and will doubtless continue for some time to come. What is sad is that, in ignoring and then distorting information about female sexuality, science has not just failed in an intellectual exercise: women have been its victims.

The momentous revelation by Masters and Johnson (in 1966) that the vaginal orgasm is a myth is probably still the most significant contribution that has been made by science to the cause of breaking down that ignorance. For that reason it is worth recapping. They described four phases of arousal in women. Excitement can be invoked by a whole variety of associations or touch; the walls of our vaginas may become lubricated, our nipples become erect and our clitorises become engorged with blood and swollen. In the plateau phase the whole of the labia and bottom third of our vagina also become engorged, which may eventually lead to the downward muscular contractions (one per second) we know as orgasm. After sexual stimulation is removed, the sensitivity of the swollen and reddened areas gradually subsides (the refractory phase). We are often able to orgasm several times fairly soon after this stage.[14]

This will probably sound very 'clinical' to most women readers. A word of warning about the 'obsessive orgasm' is therefore due. Scientists studying both animal and human sexuality have concentrated almost exclusively on a rather narrow view of sex as coital sex (fucking), in which the female's arousal is automatically considered a response to the male and is measured purely quantitatively (in terms of physiological changes like flushing, respiration rate changes). If I concentrate on the orgasm, then, it is not because I consider it the be-all-and-end-all of our sexuality (or of men's, for that matter), but simply because that is where the work has been done.

Women's sexual apparatus, in particular the clitoris, has now been closely studied, and given a higher standing(!) As Kinsey noted:

> The labia minora and the vestibule of the vagina provide more extensive sensitive areas in the female than are to be found in any homologous [equivalent] structure of the male. Any advantage which the larger size of the male phallus may provide is equalled or surpassed by the [greater area of sensitivity] in the female genitalia.[15]

Nevertheless, Freud and his followers have managed to get away with dismissing the clitoral orgasm as the 'immature' form of the 'adult' vaginal orgasm. Freud's theories served to devalue orgasms that didn't 'happen' during intercourse. Women who either didn't 'come' during intercourse or chose to have sex without men were, and still are, made to feel abnormal and guilty. Masters and Johnson, on the other hand, understood that clitoral stimulation was essential for any orgasm to take place. Once again though, they considered women who were capable of orgasm, but didn't 'come' during intercourse, to exhibit what they called 'secondary sexual dysfunction'. They insisted – and this notion was popularised – that the movement of the hood of the clitoris caused by thrusting was adequate stimulation. As Shere Hite pointed out in her exhaustive survey of American women's sexual experiences, this is almost as likely to be effective as pulling on a man's balls to make him ejaculate![16]

Hite's figures are consistent with other studies here: 10 per cent of all women have never orgasmed (and of that figure only a small proportion ever masturbate) and only 30 per cent of all women are capable of regular orgasm during intercourse with no direct clitoral stimulation. Seymour Fisher found in 1972 that if presented with an either/or choice, 64 per cent of all women would choose only clitoral stimulation, and 36 per cent only vaginal stimulation.[17]

> What Kinsey did report was a very simple fact that tens of millions of women know from their own experience: [intercourse] is less likely than masturbation to terminate in orgasm – and for some women it always or almost always terminates without orgasm.

For women, *The Hite Report* has done what no amount of detailed quantitative physiological measurements or blind faith in 'coital orgasm' could do. It personalises our own sexual responses for us. In sharing the actual words women wrote in response to her questionnaires, she reveals a whole spectrum of feelings with which millions of women have been able to identify – witness the coverage of the report in *all* of the women's magazines, and the hostility it aroused in many quarters.

Not only is clitoral stimulation essential, but orgasms with and without penetration (penile or otherwise) are *qualitatively* different for women. Orgasm during intercourse is more diffuse and deeper, often more satisfying because a partner is involved; orgasm with no penetration is more intense, 'tingly', 'piercing'. Many women maintain that some kind of penetration nevertheless adds a certain 'je ne sais quoi' (something extra; literally translated, 'I don't know what'). And, by the same token, many describe the break of flow or concentration that penile penetration brings, particularly if clitoral stimulation suddenly stops. The vast majority of women are simply not satisfied with the assumption that 'wham bam thank you ma'am' is the answer to their sexuality. Women who do orgasm during intercourse do so by knowing 'what makes them tick' and either masturbating as well or being active so as to ensure that they get the right kind of pressure on their clitorises.[18]

We now know that women are capable of coming to orgasm just as fast and effectively as men. The myth that women take longer to 'come' appears to be based on ignorance in both men and women, or on the fact that women are too shy to tell or to do for themselves what they know they will enjoy. The idea that we have all internalised at some point, that we should 'come' naturally through intercourse, and that our sexual pleasure is there to be 'given' by the 'right' man is tantamount to accepting the passive role that Freud, Masters and Johnson, and the rat psychologists[19] have bequeathed us.

Science has at least, uncovered *some* of the ignorance surrounding our bodily responses. What *The Hite Report* and other survey studies have done is to convey some 'essence' of women's sexuality that all the physiological variables mask. For women, certainly, hearing the experiences of other women is refreshing, if not reassuring. The distinctions we feel are to do with a whole range of differing sensations and emotions; how we feel about the man or woman beside us or about just being alone; how relaxed we feel and what our expectations are. The 'scientific' research, however, remains based on unquestioned assumptions – significantly that women's sexuality is – or if it isn't, should be – geared to heterosexual intercourse culminating in male orgasm.

THE MISFIT: WOMEN'S SEXUAL PEAKS AND MENSTRUATION

Not surprisingly, scientists have looked for a fit between the significant role of hormones in animals' sexual arousal and peaks of sexual interest in the human female. Before 1920, menstruation in women was thought to be the same as oestrus in animals. Observers like Havelock Ellis (1910) were therefore not surprised to find that women reported a peak of actual desire (often accompanied by erotic dreams) just before, just after, or all the way through their periods. 'It is at this period that masturbation may take place in women who at other times have no strong autoerotic impulse.' Kinsey had fourteen cross-cultural studies to back up his conclusion that: 'Evidently the human female in the course of evolution has departed from her mammalian ancestors and developed new characteristics which have reflected the period of maximal sexual arousal near the time of menstruation.'[20]

Despite all this, scientists would still be much happier (certainly if my undergraduate lecturers were anything to go by) if our sexual peaks coincided with ovulation (fourteen days after menstruation starts), as they do in animals. The so-called 'paramenstrual peak' is difficult to explain. Some go as far as to suggest that it has actually disappeared, and maintain that people copulate more at ovulation because they 'know' it to be more productive. This is as may be, but women have continued to report a sexual peak around their periods. McCance, Fisher and others have since confirmed the existence of two peaks. Figures vary. R. D. Hart found that 59 per cent of his survey reported a 'paramenstrual peak'. Sherfey and Kinsey felt that 90 per cent of women are most sexually active right after their periods. The possibility they had all failed to consider was, once again, that the two peaks are actually qualitatively different for women.[21] Money and Ehrhardt have now addressed themselves to this aspect. At ovulation women report a desire to be 'occupied'. They feel more 'receptive', in Beach's sense of the word. Around menstruation the desire is more initiatory; women are more active. This would fit in with McCance's other finding that single women tended not to report a mid-cycle peak while married women do. As Shuttle and Redgrove commented in their book *The Wise Wound*, 'This could be disturbing news to men reared on the idea that it is the male prerogative to initiate sex. The combination of bleeding and increased sexual capacity is a formidable one to the conventional view.'[22]

The problems facing a science that tries to quantify our sexuality are summed up very well in Dr Mary Jane Sherfey's explanation of women's sexual capacity around menstruation. In clinical practice, she recorded incidences of multiple orgasmic sexuality owing to premenstrual congestion of blood in the genitals. Women who have achieved up to fifty orgasms in one session with the use of a vibrator she describes as 'cases of nymphomania without promiscuity' – which she considers a cultural statement: 'from the standpoint of normal physiological functioning, these women exhibit a healthy, uninhibited sexuality – and the number of orgasms obtained are a measure of the human female's orgasmic potential'. She quotes Masters:

'A woman will usually be satisfied with three to five orgasms. . . .' I believe it would rarely be said, 'A man will usually be satisfied with three to five ejaculations'; the man is satisfied. The woman usually wills herself to be satisfied because she is simply unaware of the extent of her orgasmic capacity. However, I predict that this will come as no great shock to many women who consciously realise or intuitively sense their lack of satiation.[23]

It is revealing that Sherfey considers women to be 'oversexed' by cultural standards. Certainly if one compares, as she does, our respective capacity to orgasm, and if that is the *cultural* measure of an individual's sexuality, then the statement is true. We can only conclude that this cultural measure is a male standard. Sherfey herself resigns womanhood to a life of sexual containment for apparently biological reasons: at menstruation, she claims, orgasm leads to greater congestion and therefore greater desire, leaving women unsatiated. However correct or incorrect this hypothesis is biologically, I suspect that few of us would accept a cultural definition of this kind, which measures sexuality purely in terms of 'orgasmic potential'. It has little to do with the vast spectrum of experiences and emotions that make up our sex-lives. Rather, I would suggest that what is needed is a culture that recognises the *qualitative* differences in woman's sexuality, and values those differences in their own right. For that reason I want to go on to consider how culture generally has expressed and mediated that 'otherness' of women.

THE CULTURAL AND SEXUALITY

In other, maybe less 'hung-up', cultures, sex assumes a wider definition than that of 'coital orgasm'. The result is, usually, that women take a more active role and can thus express their own arousal. In the South West Pacific Islands, for example, sexual enjoyment is considered the key to marriage. Heterosexual behaviour embodies a respect for both female and male sexuality. In our terms it is more feminine, with long periods of foreplay including prolonged genital stimulation. Penetration usually lasts only fifteen to thirty seconds. The woman who is unable to reach orgasm this way is unheard of. Either wife or husband can break off the marriage if intercourse is infrequent (approximately less than once in ten days). Similarly, in Trobriander society, the word 'orgasm' actually means 'seminal fluid discharge' – for both the woman and the man. Malinowski found women there to be more assertive, vigorous and active than in the West, and often more than men there. ('Active' in Trobriander society is measured by the extent of scratching on the partner.) By contrast, the Arapesh people in New Guinea tend to devalue sex. Marriage is arranged before puberty and is valued solely for purposes of childrearing. Interestingly enough, it is expected that the girls, rather than the boys, will be the first to 'feel inclined'. [24]

Woman as sexual initiator is not new to cross-cultural studies – in some societies they are exclusively so. Ford and Beach elucidate some of the implications for our own culture:

> In contrast [with women's submissiveness and modesty in the West], the societies which permit or encourage early sex play usually allow females a greater degree of freedom in seeking sexual contacts. Under such circumstances, the sexual performance of the mature woman seems to be

characterised by a certain degree of aggression, to include definite and vigorous activity, and to result regularly in complete and satisfactory orgasm.[25]

It is surely significant to an understanding of our own sexuality in the West to note that only in societies where female sexual desire is considered socially dangerous, as in the Manus, is it actively curbed.[26]

The menstrual taboo has been well documented – in this book and elsewhere (see Chapter 5). It may come as no surprise that this custom is now beginning to be viewed as a response to the 'dangers' of women's sexual desire at that time. The menstrual taboo, in its various forms, has affected both the self-exclusion of men and the sexual isolation of women. It may have resulted from a conscious choice by women, or it may have been rationalised as a necessary response to the very different pressures that have existed on the survival of early human societies.

> Attitudes taken by different societies towards menstruation are rationalised in various ways, but the reasons advanced almost always are based upon the attitudes of the woman's masculine partner or other associates: rarely do they seem to arise from any recognized alteration in her own desires or tendencies.[27]

If it was a male invention, was it because of an aversion to our blood on behalf of our brothers, or was that just a symptom of a deeper fear of our sexual prowess at menstruation? Certainly, neither our supposed discomfort at the sight of blood nor the 'survival' pressures of twentieth-century life seem adequate explanations of the mystique still surrounding our periods. Indeed, it has even been suggested (by Redgrove and Shuttle) that premenstrual tension, as we know it, may be a behavioural and social expression of the paramenstrual sexual frustration to which Sherfey referred.[28]

Evolution has endowed us with the potential for a vast range of sexual experiences that go far beyond the limitations of heterosexual intercourse; masturbation, as we have seen, does take place in 'lower' mammals (even if the extent of excitement is not recognised). In the societies described earlier, masturbation in children and adolescents is expected if not condoned. Masturbation in adulthood appears to be universally regarded as a substitute for partner sex:[29] in the South West Pacific Islands, it is simply considered 'silly', whereas our society generally considers it to be perverted and has punished women who masturbated, particularly in the last century.

Most anthropological literature insists that masturbation in men is far more common, although the Lesu people of New Ireland expect women to masturbate in the absence of a partner.[30] Azande women, coming from a culture of great phallic symbolism, use a wooden root for penetration, but if caught they are beaten by their husbands.[31] In our own society, approximately 40 per cent of all women at the age of forty have masturbated as compared with well over 90 per cent of all men. Ann Oakley makes the very pertinent point that, if we consider the social sanctions against girls and women exploring their own sexuality early in life and the decreasing incidence of masturbation in men after the age of eighteen or twenty (regardless of marital status), then women may masturbate at least as much as men.[32] It should of course not be forgotten that masturbation, though more generally accepted as a male activity, is possibly considered more shameful in men just because of the associations of not having a (female) sexual partner. It may be true to say that, although fewer women do masturbate, either because they don't know how or because they feel guilty, those who do are more able to enjoy it freely.

That women masturbate 'instead' of having sex with a partner need not be devalued. For people in all cultures, force of circumstance often provides little alternative, and in any case, for some women celibacy is a positive and conscious choice not to have sexual relationships as such with men or women.

The ability of all humans to become sexually aroused with members of the same sex is another heritage of our evolution. Homosexual behaviour, though not exclusive homosexuality, is common to all animals. Humans, like other animals, have to learn the techniques of heterosexual intercourse. On this basis, homosexuality is really not so surprising as some would have us believe. Again, we find that female homosexuality has not been recorded as much as male homosexuality. Ford and Beach found that forty-nine out of seventy-six societies studied condoned homosexuality, but of these, only seventeen indicated lesbianism specifically. In many societies, including our own, homosexuality is believed to be linked to a gender identity with the opposite sex (see Chapter 6). Indeed, where homosexual partners 'marry' these identities have become institutionalised. Mohave Indians always regard one of the partners as being of the opposite gender. This is sometimes reflected right down to their assuming the 'appropriate' sexual behaviour, be it active or passive.[33]

Homosexuality, then, is certainly not biologically 'unnatural'. Society and its cultural norms appear to influence deeply the 'incidence' of homosexuality as well as the form of homosexual behaviour. The associations of 'femininity' and 'masculinity' with passivity and activity, found even within homosexual relationships, reflect the enormous extent to which our sexuality is socially defined. In our society we tend to expect to find sexual and emotional happiness within monogamous, heterosexual marriage. This expectation probably explains in part why more women do not enjoy lesbian relationships – either as well as or instead of heterosexual relationships.

It seems to me likely that we are all essentially 'bisexual' in psychological makeup. We could envisage a day when freer sexual expression revealed a plurality of sexual preferences, from exclusive homosexuality through bisexuality to exclusive heterosexuality. Moving in that direction involves realising and developing the masculine as well as the feminine within us: so-called androgyny.

The question of how much of our sexuality is culturally and how much is biologically defined is not easily answered. Yet it is important if we want in any way to branch out of existing definitions of our sexuality. It is a truism to state that the individual is affected by society and in turn acts upon society, and that both respond to and impinge upon the environment. It is still useful to remember that humans, possibly more than any other species, are conscious beings, aware of their potential to act and so change these relationships. Society will always shape our sense of our selves as sexual beings: our biological sex 'urge' assumes a cultural expression in which sex is institutionalised, and the dominant sexual ideology defines our sexual experiences to·a large extent. The demands of a hostile environment on early societies may well have necessitated institutions as marriage or the menstrual taboo, but if this is so, it seems unlikely that the natural environment of today is the reason for their continued existence.

Before going on to consider the institutions and the ideologies of the last two centuries of our society, there are a few unanswered questions on the subject of our evolution that might shed some light on this 'problem'.

EVOLUTION: THE UNANSWERED QUESTIONS

From our glance at animal and human societies, two things at least emerge. First, although the 'obsessive orgasm' pervades all inquiry, it represents only one small part of sexual activity and experience for both women and men. Second, this consideration aside, there is no logical reason for insisting that our orgasms result 'naturally' from intercourse. Clearly, for the evolutionist the question of how and why sexuality and specifically female sexuality evolved is problematic.

Oestrus in primates ensures that copulation is predominantly fertile. The female will often reject male sexual interest and, for the anoestrus female at least, rape is virtually unknown.

> The adaptive value of such synchrony is so obvious that its disappearance during human evolution is perplexing, and it is apparent that compensatory modifications had to occur which would ensure that a sufficient number of matings would take place when females were about to ovulate and thus could be pregnated.[34]

As Beach points out, *one* solution would be to increase the overall frequency of copulation. This would require males to find copulation attractive at all times (without having to rely on 'suitable' hormone levels in the female), and for the female to experience sexual desire and gratification. Women have certainly evolved permanent secondary sexual characteristics, such as breasts, which are not present in primates, but are universally erotic (to both men and women, I suspect) in human cultures. Similarly, women crossculturally are able to convey sexual desire and availability. 'It is, however, interesting to note that sources of reward for the copulating female have not been dealt with by evolutionary theories in general.'[35]

One could also add that, if Beach and others were not quite so dismissive of the possibility of female orgasm in other animals, they might find it easier to understand just how the human female orgasm evolved.[36] The question of why the clitoris, as an organ whose sole function appears to be sexual pleasure, evolved still remains a problem for scientists. Unlike the majority of primates, the most universally common position in humans is front to front, which, at least if the woman is on top, increases the possibility of female orgasm during copulation.

Obviously, only male orgasm is necessary to 'successful' – i.e. reproductive copulation. Furthermore, it has often been noted that impregnation may be more likely if the woman has *not* orgasmed to the point of decongestion. On the other hand the downward contractions associated with female orgasm may help to 'bring on' male ejaculation.[37] If this is the evolutionary function of the female orgasm, it may explain such features of female arousal as vaginal 'sweating' (or lubrication) and the 'vaginal ache', associated with the desire for penetration of some sort, that many women spoke of in *The Hite Report*.

Physiology aside, many scientists have turned their attention to the possible role of the nuclear family as providing the opportunity for regular and frequent intercourse, by ensuring relatively permanent pairbonding. It is when we turn to these social aspects of human sexual behaviour that scientists fall into the murky waters of evolutionary speculation, and biological determinism is confronted with political questions.

For a long while, we were treated to the 'Ever Ready Vagina' theory. Briefly, this theory postulates that: (1) the human female is 'constantly receptive'; (2) 'the human male is constantly potent and needful of sexual relief through copulation'; (3) the nuclear family occurred *because* 'the females' unvarying receptivity gratified the males' perpetual demands'. The assumptions are oceanic! First, it denies all that is already understood about the role of behaviour as a necessary prelude to copulation in animals. And second, it makes a nonsense of everyday human experience. As Beach wittily points out, 'no human female is "constantly receptive". (Any male who entertains this illusion must be a very old man with a short memory or a very young man due for a bitter disappointment.)'[38] I doubt whether all men walk around permanently needful of sexual relief; and it is certainly not true that sexual release *has* to be found in intercourse.

As for the third point, that the nuclear family evolved to ensure frequent copulation, E. O. Wilson for one would disagree. He maintains that the 'physical pleasure of the sex act' alone serves adequately to ensure the continuation of the species. He also is unconcerned with explaining the source of reward to the copulating female (though he claims the right to assert that 'they [women] never attain the peak of readiness that defines the oestrus in other animals'. How can he possibly know?!) To him, pairbonding and the nuclear family serve a wider function related to the sexual division of labour around childrearing and food provision. He argues the whole thing the other way around – that sexual interest and readiness serve to cement the pairbond, which itself is the key to the day-to-day economic survival of human societies. To him, then, sexual preoccupation has more to do in evolutionary terms with pairbonding than with reproduction. (That it also 'helps' to reduce the potential for male 'aggression' and 'competition' is also useful to his schema.[39])

The argument goes on to 'explain' the sexual gender differences that have become associated with femaleness and maleness. In Beach's words, 'sexual behaviour reinforces family structure and family structure reinforces sexual behaviour'. It is quite possible that there was some sexual division of labour in early human societies (see Chapter 3). We can safely assume that reproductive as well as productive pressures have played a role in shaping family structures, and that these in turn helped define the norms for women's and men's sexuality. The political explosions occur however over the question of just how much these differences in sexual behaviour have become 'innate' and how much they are socially defined. Those who claim that they are predominantly 'innate' conveniently defy others to confront social definitions.

Whatever the answer to the 'nature–nurture' controversy, it is clear that society now enshrines heterosexual monogamous intercourse as *the* basic sex act. Feminists have quite rightly asked questions about why human societies left the early days of shared childrearing and 'loose' pairing for the confines of the nuclear family. The emergence of the nuclear family gave man a very significant prerogative; property rights, taking the form of insistence on monogamy and virgin marriages, bound child to (biological) father and woman to husband. This kind of social organisation was bound to carry standards of what could be considered 'acceptable' sexual behaviour.[40] Some have suggested that these developments reflected a male need to establish parenthood on the basis that 'the only way a man can be absolutely sure that he is the one to have contributed that sperm is to control the sexuality of the woman' – man has no hold on the process after conception. Certainly if monogamy was required of woman, man has found numerous ways of ensuring it throughout history:

> He may keep her separate from any other man as in a harem, he may threaten her with violence if she strays, he may devise a mechanical method of preventing intercourse like a chastity belt, he may remove her clitoris to decrease her erotic impulses, *or* he may convince her that sex is the same thing as love and if she has sexual relations with anyone else, she is violating the sacred ethics of love. This last method is the one used most commonly in the United States today.[41]

Whatever role one believes that pairbonding fulfils, it is illogical for evolutionists to suggest that female enjoyment of sex 'arrived' in order to cement pairbonding. Female sexuality in itself guarantees nothing by way of faithfulness to one man. In fact it is more likely to have the opposite effect.

Through enormous sections of women's history, women's sexuality has been controlled by physical mutilation and containment. When it comes to our own, more recent, history the medical profession –

the applied wing of science – has probably played the largest part in that control. In the nineteenth century it served to bolster up the prevailing Puritan ethic, by its practice and by its theory. It is responsible for many of our present-day sexual 'hang-ups', which the rapidly growing field of sex therapy is now trying to answer.

NINETEENTH-CENTURY MORALISM AND THE MEDICAL PROFESSION

Puritanism, in its various forms, can be characterised as the 'hard work ethic'. Man, so it went, was at core base, just a savage; only diligence and 'rationality' would raise him (and thus society) from his sordid moral state. In this climate science became the answer to all ills: 'Science', claimed a prominent gynaecologist, 'in its future progress may yet refine us and separate the pure from the impure. . . . Health is the great prophylactic against sin. . . . The vigour of the body is the basis for a robust morality.'[42] Sex, for both women and men, unless specifically directed towards procreation, was wasted energy and was therefore damaging to the health. Anything other than the 'missionary position' (that is, face to face with the woman on her back) was considered harmful because, it was thought, fertilisation might be impaired. Women of the nineteenth century were exalted as reproducers. The female orgasm was peripheral to conception, and so women were thought (and expected, in the case of middle-class women) to be 'naturally frigid'. Instead of sexual gratification, women 'naturally' derived their deepest, almost religious, satisfaction from motherhood. Their role as moral enclaves for the sons of tomorrow was omnipresent.

Predictably, the Victorians' definition of woman's sexuality ran into problems with nature. On the one hand, she was more spiritual, sacred and untouchable. She did not 'need' sexual release like her brothers, who were granted the double standard for their weaker, more base, animal moments. On the other hand, woman herself was more animal and less endowed with the superior intellectual capacity of man. Man had to guard himself against the fearful 'sperm-sucking propensities' of woman's sexual prowess. Lallemand explained the reason for this fear in his tome on *Conjugal Sins*: 'To the man there is the limitation of physical capability which no stimulants from within or without can goad to further excess . . . The eroticism of woman has no boundary.'[43] He would appear to agree with Sherfey, or at least to reflect her conclusions: ultimately, man as rationality embodied had a moral duty to civilisation to *contain* the 'impurity' of woman's sexuality.

Woman's role in the broader society reinforced her sexual fate. As man assumed responsibility for the 'progress' of the new era and the social imperative to be active and dominant in the public world of business, so he had to dominate in the private world of home and bed. Woman's passivity sexually reflected the passivity required of her socially – unless, of course, she was obliged to work and thereby forfeit her 'womanness'. (Even working-class women were eventually faced with the same morality.) Religion served to resolve the resulting 'war' inside woman. She was persuaded that sex was a necessary evil, endured only because of the reward of motherhood and because of her duty to be endlessly supportive and faithful to her husband. She had to guard her children from their natural instincts (to masturbate, for instance) in the vulnerable stages of their moral development. With their increased dependence on the nuclear family, and with the growth of childhood as an institution, children became vessels of purity which only vigilance, and punishment if necessary, could preserve.[44]

Alex Comfort summed up the situation well: 'Sex is tolerable at the subsistence level, lip-service is paid to its use as an expression of love, but any reciprocity of enjoyment between the sexes is suspect – it has no place in the moral gymnasium with our safety nets.'[45] Concepts of health engendered by the medical profession became a very large safety net where women were concerned. As woman was defined solely in terms of reproduction, bad health – both moral and physical – was automatically put down to dysfunctions in her reproductive organs. The role of the gynaecologist was in the 'detection, judgement and punishment' of sexual disease and 'social crime'. In this climate, pelvic surgery was simply a 'social reflex'.[46]

Clitoridectomy (the surgical removal of the clitoris) was the first operation performed to check supposed mental disorders in women. Innovated in England in 1858, it soon fell out of medical favour there, only to become popular in the United States. Castration (removal of the ovaries), which then became more widespread, was first developed in 1872. The next year, a Scottish Dr Battey advocated the operation for what he called 'non-ovarian conditions' (in particular masturbation). By the 1880s it had superseded the practice of clitoridectomy, so that by 1906 it was estimated that 150,000 American women were without ovaries.[47]

The diagnosis of 'non-ovarian conditions' was a social judgement aiming to prevent the 'unfit' from breeding and thus polluting what was called the 'body politic'. The 'unfit' included American migrants and blacks, the insane and any women who had been corrupted by masturbation, contraception and abortion. These all had to be eradicated. From the 1890s until the Second World War, mentally ill women were castrated. 'An insane woman is no more close to the body politic than a criminal.' Reproduction generally and menstruation specifically were thought to make women more prone to mental illness. 'Certain women especially at the menstrual epoch are so overcome by the intensity of sexual desires and excitement that they practically lose self-control and all modesty.'[48] Logically then, woman was unhealthy unless pregnant, and her crime was to have expressed her sexuality.

The medical men of the time assumed all responsibility for making the moral decisions required to save women in these times of supposed suffering. Dr E. W. Cushing of Boston had a totally clear conscience: 'Orgasm was disease and cure was its destruction.' His sense of purpose was clear: 'Patients are improved, some of them cured . . . the moral sense of the patient is elevated . . . she becomes tractable, orderly, industrious and cleanly.' Middle-class women in particular had so internalised these values that the gynaecologists were 'answering their prayers'. One of Cushing's patients, relieved by his scalpel of the 'temptation' to masturbate, wrote, 'A window has been opened in heaven [for me]. . . .' It was, of course, not working-class women from whom the gynaecologists made their money. Those women either did not have the time to consider their 'impurity' or, when they could afford it their 'symptoms' were not cured by castration because, in medical opinion, they did not convalesce sufficiently for the operation to succeed.[49]

Medical opponents of the operation criticised it not on moral grounds, but because in their experience surgery often failed to return women to their 'normal' role.[50] There was controversy as to exactly what it was that the operation effected. Dr Warner in fact maintained that any changes of behaviour that resulted from castration were purely psychological. This did not lead him to abandon the operation and question woman's psychology, however. Having decided that 'the brain was involved in nymphomania, not the ovaries', Dr Syminton-Brown concluded that the operation was still valid because it worked by 'shock effect'.[51] (The use of brain surgery such as lobotomy for sexual 'disorders' appears to be the logical twentieth-century extension of this argument.)

Confusion reigned among the medical profession towards the end of the nineteenth century when thousands of so-called self-styled city women, many of them suffragettes, started to *choose* sterilisation. For these women, sexual desire became if anything more pronounced after castration, and so came to be seen as a symptom of the operation rather than the reason for it. Doctors were divided between those who believed that such women should not be allowed to breed and so performed the operation willingly, and those who felt threatened because women were choosing *not* to fulfil their reproductive role.[52]

It is fairly clear that these definitions of our sexuality based solely on our reproductive potential not only landed women with more children than they might have wanted, but also denied whole generations of women self-respect as sexual beings. Medical theory and practice played a major part in giving credibility to Victorian values. Its definitions of illness obscured sex and sexuality for both the female patient and the male practitioner. Doubtless, twentieth-century medicine has progressed somewhat; gynaecology is no longer seen as the only area of health that is relevant to women, although some branches of medicine – in particular psychiatry – tend to treat women as a 'special case'; the new field of sex therapy at least recognises women as sexual beings. However, before considering the role of science and medicine in this, the age of sexual liberation, we must first look at how the values surrounding women and our sexuality have changed.

LIBERATION AND THE NEW SEX THERAPY

'Give me back my past, my childhood, my body, my life!'

This graffiti, originating in the passionate turbulence of 1968,[53] is a testimony to the extent that we are all distorted and confined by the dominant social values and institutions. A science colleague admitted to me recently that he was only just beginning to realise that women have sexual fantasies of their own. Woman is no longer considered 'naturally frigid', but the images we find in the media seem to perpetuate the idea that man is more needful of sexual 'release' than woman and that woman's sexuality is geared to fulfilling that need; soft porn unfolds woman, the ultimate sex object, who silently and facelessly satiates the almost burning, conquering lust of man's gun-like organ. Romance on the other hand reveals woman, the Hemingway-like ideal, who 'comes' just as he does and thus consummates 'true love'. Woman, it seems, is either 'nice' or a 'cunt'.

This statement is not really as shocking as it might appear. Our sexual experiences are as much socially formed as they were a hundred years ago. Women have felt guilty if they don't give their man sexual release, and men also have expected that 'release' to answer women's sexual needs. Hence the proverbial question, 'Did you come?', and the complaint of many women that they can never really say 'no'. In this, the supposed age of liberation, the idea of 'nice' versus 'cunt' has undergone a subtle change. 'Nice' women don't necessarily have to keep themselves for 'Mr Right', but they can't afford to be too forward either; in bed the sequence has, I suspect, not changed enormously. Not only do we *expect* men to be more active and assertive, but often we adulate those characteristics in a male partner ('he's a *real* man'). Expressing our own sexual feelings is either threatening or not 'nice'.

The split between femaleness and maleness is about more than just passivity versus activity. Woman is supposed to bring emotion to what for man is supposed to be a purely physical act. She is more sensual and man is more sexual in the sense of craving orgasm, while in real life these splits rarely exist in extreme, for our attitude to sex seems to be a combination of both our social relations as women to men and the age-old mystique surrounding our bodies. As Sheila Rowbotham comments, 'The sexual distinction at once binds us more closely to our oppressors and distinguishes us more sharply from them . . . after all, a man enters us through our vagina; we perceive his body through our sensations of him on top or underneath, inside and outside.'[54]

As women, we carry the image of ourselves as receptacles of male sperm into the deeper parts of our personalities. The 'Ever Ready Vagina' makes us not only physically but also psychologically incapable of rape, though always capable of being raped and therefore fearful: 'Along with the males' greater aggression in other fields goes his aggression in the sphere of sexuality: males initiate sexual contact, and take the symbolically if not actually aggressive step of penetration – a feat which is possible even with a frigid mate . . .'[55]

Fear of rape in a sense epitomises the 'privateness we feel for our bodies and in particular our pelves. Seymour Fisher, in concluding his book *Understanding the Female Orgasm*, concedes, refreshingly, that the medical profession must take some responsibility for this:

There are few aspects of modern women which have aroused as many nurturant efforts on her as her reproductive system. The gynaecologist has assumed an extremely onerous rescue role. It is my hunch that much of the anxiety which has been focused on the woman's reproductive system is a coded and condensed way of atoning for other unfair things that are generally done to her.[56]

He goes on to trace the 'problem' of inhibited orgasming to 'the fact that little girls get innumerable messages which tell her [*sic*] that the female cannot survive as a single entity and is likely to get into serious trouble if not supported by a strong and capable man.' In this situation, sex as loving happens, if anything, in spite of rather than because of our concepts of love and sex; the idea that 'all you need is love' has probably held back even the more courageous women from asking for 'different' in bed.

Sexual morality has of course often made recourse to arguments of health and has thus further separated sex from loving. In the nineteenth century doctors claimed that only women could pass on gonorrhoea and that they were most likely to get infected around menstruation. This was used as yet another argument to dissuade men from having intercourse with menstruating women.[57] Today, parents warn their children against the horrors of sexually trans-mitted diseases, often without explaining what these diseases are and how they can be avoided or treated. The diseases remain mysterious, and people venture into sex in fear of inordinate punishment rather than gratification.

Clearly the subject of sex, if not sexuality, is much more open nowadays. It is ironic that, when the work of Masters and Johnson was first published, they were accused of taking the 'romance' out of sex by making it sound so mechanical; demystifying women's sexuality involved pointing out that women do not need to be 'in love' to enjoy sex. Great steps have since been made towards sex education in the broadest sense. Popular women's magazines are beginning to talk about sex – so long as it is heterosexual sex – and in the United States courses are being run on the subject, with textbooks and journals to back them up. Individuals and couples are more willing to admit that they are unhappy sexually and sex therapy has now become respectable.

In practice, however, many of these advances reflect the narrowness and illusions that have plagued so much of the scientific research. While it is now considered to be terribly 'liberated' to recognise women's sexual needs, what is often offered is a 'male' version. Progressive sex books, for example, are generally orientated to male orgasm; sex is described as foreplay, penetration and thrusting. The notion of 'foreplay' is interesting here. It is often seen as an essential prerequisite to women enjoying intercourse, but is rarely seen as an end in itself. Sex stereotyping is perhaps worst in the popular press, though it continues to draw credibility from the scientific community. A series of articles, for example, run in a British daily newspaper recently featured the results of a survey run by therapists (including Masters and Johnson) under the title 'Beware the Virgin Husbands!'. Implicit in it was the notion that it was really far more serious for the man to be sexually inexperienced than the woman; men who were still virgins up to or after their wedding nights tended to be rather effeminate anyway – they didn't 'go out drinking with the boys'.

For the scientist, clinical experience in sex therapy has become the major source of research material. In a situation where people will reflect the 'obsessions' of their own society, it is easy to see how science is often guilty of 'manufacturing' syndromes. Inability to achieve orgasm during intercourse is the most common 'complaint' of women who visit clinics. Helen Kaplan, author of *The New Sex Therapy*, comments: 'These impressions are in sharp contrast to the view held by many experts, and shared by the general public, that coital orgasm is the only *normal* form of female sexual expression and that orgasm attained primarily by direct clitoral stimulation is somehow *pathological*.'[58]

The tradition of Freudian psychology must take much of the responsibility for this particular obsession. Freud maintained that, owing to the sexual feelings associated with the mother and the father in their family, adolescents exhibit bisexual leanings, which in the woman is manifest in her enjoyment of clitoral orgasms. In order to 'graduate' to vaginal orgasms, women have to muster enough 'courage and self-sufficiency' to break off those family sexual associations.[59] The idea is rather like the attempted triumph of mind over body, so symptomatic of the nineteenth century: psychoanalysis is largely geared to 'helping' women to 'mature' in such a way that they accept that only their husbands could (or should?) make them have orgasms. Is it so very *rational* for us to want a 'mature', 'normal' sex life?!

What role, then, does science have to play? Clearly, it has been valuable in demystifying women's sexual anatomy, although reference to the clitoris as the analogue of the penis is not necessarily helpful to those striving towards a sense of their own womanness. More recently, sex therapists have tried to remove the blame for our 'failures'; premature ejaculation in men is now identified as a major barrier to women's arousal in intercourse and has, subsequently, been classified a *male* 'dysfunction'. Again, such statements are not particularly helpful to men any more than they are to women. If sex therapy is to be useful it must get away from further manufacturing 'guilt syndromes'. It could start, at least, by beginning to see sexuality as being an integral part, an 'essence', of every individual. Sexual gratification is important, but it need not be measured in terms of 'achievement', as though sex were something external which, like so many other things in our society, we have to compete for by following the right set of rules or, in this case, acrobatics! Sexual self-definition means defining what you want and what you feel. In many respects women are better 'suited' to do that at the moment than men. If only there were a value on our womanness!

It is clear that there is a gap between what science has deemed to be our 'normal' sexual experiences and what women know about their sexual feelings. Such a gap cannot be veiled for ever by the practices of normative medicine. There is no disputing that our sexuality has been controlled both by individual men and by pervasive social institutions. And yet, I am optimistic enough to believe that society is ready for a change of outlook. It will require very far-reaching changes both in how women and men view their own sexual selves, and in the expectations we have of one another; misogyny and male definitions of women's sexuality are based on a history of ignorance and fear – a history characterised by a dual egocentricity and insecurity in men. Women are gaining the courage to confront this insecurity in their personal lives. Gradually we will be able to identify the 'otherness' that is common to all women's experience – both sexual and social.

Inevitably, society will not pass unchanged through this process. We may gain sexual happiness individually and in spite of society, but sexual self-respect for women will come only through changes in society. It will take the voices of many more to redress the balance set by centuries of fear and denial. We still have to speak out, however, against the media industries that make money out of objectifying our bodies and our sexuality; we have to speak out against the very narrow view of heterosexual intercourse and gender which they portray and which permeates all the scientific research. We need a broader, more open, definition of sex, one that defies the 'compartmentalisation' of Western science and recognises the flow within us all between sensuality and sex, femininity and masculinity; one that allows everyone the freedom to establish her uniqueness and preciousness as a sexual being. The most positive thing we women have retrieved from the nineteenth century is that sex is about enjoying ourselves. Let us get on with it.

Part Three

TECHNOLOGICAL CONTROL

8

Technology in the Lying-in-Room

The Brighton Women and Science Group

Shortly after the Bolshevik Revolution of 1917, the Soviet Government issued a decree on infant mortality, saying: 'In Russia two million infant lives which had hardly been kindled were annually extinguished because of the ignorance of the oppressed people and the inertness and indifference of the bourgeois state.'[1] Nor was it any different in Western Europe, where there was always a high risk of infant or maternal death. Over the two centuries preceding this, urbanisation increased as the Industrial Revolution proceeded. With it came a growth of hospitals specifically designed for 'lying-in', for women in labour and just after – or at least, for those women of the working class who were too poor to be attended privately at home. The filth and poverty of those hospitals mirrored the filth and poverty of the working and housing conditions of the same poor.[2] Deaths, under these conditions, from puerperal sepsis (childbed fever), caused by attendants with filthy hands touching the labouring women, were common. Deaths were so frequent in some hospitals that pregnant women would beg not to have to go into them.

Famous for her nursing zeal in the Crimean War, it is less commonly known that Florence Nightingale spent most of her life attempting to institute better hospitals, with better organisation and staffing. One of her activities was to institute a study of puerperal fever after it broke out in a training school for midwives which she had herself initiated. She was horrified to find that, in nearly all the major hospitals with labour wards, it was rare to find accurate records of deaths or of the causes of deaths, and her efforts to discover some facts were met with much suspicion by the medical men. Eventually she pieced together the story, to discover that the risk was much magnified when women were together in a ward, and was minimal when they gave birth at home. She wrote: 'Not a single lying-in woman should ever pass the doors of a general hospital. Is not any risk which can be incurred [by staying at home for delivery] infinitely smaller?'[3]

In 1861 a Viennese physician, Semmelweis, published a book detailing similar findings. After watching the agonised death of so many women, he slowly realised that the male doctors who were attending the labours had often come straight from the post-mortem room.[4,5] He pleaded for greater cleanliness among such doctors, but at first his plea was ignored. With so many grotesque deaths on his conscience he, ironically, died of a similar germ-carried disease in 1865. One year later, in 1866, the last cholera epidemic ravaged Britain, taking 14,000 lives, catalysing rapid parliamentary reforms in public health, and paving the way for a greater understanding of the cause and prevention of germ-carried disease.[6]

Medical care in Western, industrialised countries today concentrates on personal health care and on curative, rather than preventive, medicine. This has led to a concentration on large, centralised units with many technological facilities at their disposal – the large county hospital has become the centre of new techniques in medical science. Childbirth has changed more and more rapidly over the years, from being a personal, all-woman, shared event to being a pathological state from which women will recover, under correct supervision and care. The desire to create better, safer facilities for labouring women has turned into a system whereby women are processed through pregnancy and after, according to some 'average' pattern. Childbirth has become another illness for which we need treatment and cure, rather than a personally and socially meaningful event. This chapter will deal mainly with the results of the change in ideas towards birth and labouring women – in particular, the effect technology is having on the process of birth.

The original reasons for encouraging hospital deliveries arose from a desire to reduce the maternal and child mortality rates. However, in recent years there has been a fresh explosion of argument on the necessity for 100 per cent hospital births, as well as protests against the treatment received in many hospitals.[7] In some ways, this is a repeat of the 'hospital v. home confinement' arguments, but progress in medical science and technology has changed a simple question of choice into a fundamental question about hospital organisation and resources, the politics of health care, and the desires of the people involved.

The decision (embodied in the 1970 Peel Report[8]) to aim for 100 per cent hospital deliveries was preceded by a long wrangle over the professionalisation of medicine and obstetrics and, in particular, the reduction in status and satisfaction of midwifery.

Medicine only recently became a profession. There is considerable historical debate about the forces behind this move to 'professionalisation' – whether it was a desire to protect patients from unskilled doctors and quackery; whether it arose out of the sudden severe increase in mass health problems with the coming of capitalism and the growth of urban populations; or whether it had more to do with bolstering the doctor's income, prestige and authority. It was probably a combination of all these, and other, factors. One thing is, however, certain: the professionalisation of medicine had a specific effect on women – it excluded them. Women who throughout history had tended the sick were shut out by laws and examinations, and had a long battle ahead to earn the right to train as doctors.

Midwifery, so obviously a female domain, resisted the pressure of the male medical profession for a longer time, but became vulnerable with the rise of scientific medicine during the nineteenth century.[9] Scientific knowledge ousted experiential knowledge of the women midwives, and they were *denied* access to the former. The rise of science and the defeat of women at the childbed are closely linked. Sophia Jex Blake, one of the first women allowed to qualify as a doctor in Victorian England, even suggested that it was the invention of a particular technology – the forceps – and the belief that only a surgeon (that is, male) should/could use them that led to the male takeover of childbirth. The evolution of *'scientific'* medicine was the *one* way in which doctors could insist that men rather than women belonged at the childbed. In the call to 'science', the personal and subjective meaning to women of birth was despised and ignored.

Before the Second World War the majority of women gave birth at home, attended by domiciliary midwives and possibly their GPs. However, in 1944, a few years before the advent of the National Health Service, the Royal College of Obstetrics and Gynaecology advocated that 70 per cent of all births should occur in hospitals, especially of first babies and for women who might have complications.[10] As a result of this, women were increasingly persuaded that hospitals were the best places for their babies to be born, so that by 1959 only 36.6 per cent of all deliveries were at home, and by 1971 only 11.1 per cent were at home.[11] Now, approximately 90 per cent of all deliveries occur in hospitals in Britain, and it is becoming increasingly difficult for a woman who expresses a preference for a home delivery to have one.

While official policy is for 100 per cent hospital births, there has recently been a call for a return to home confinements, particularly for women who have had previous uneventful births. Although some doctors tend to over-emphasise their own role in the greater safety of childbirth today,[12] a few are beginning to recognise that home confinements – for those women who choose them – are safe.

> [Although] there must be some uncertainty about the true situation . . . two points are beyond doubt: that home confinement under present conditions in Britain is very safe for both mother and child . . . and that the difference in safety for low-risk cases between home and hospital confinements is small.[13]

Pregnancy can obviously be difficult, even dangerous, for some women, but the function of the monitoring services (GPs, ante-natal classes and clinics) is to try and determine, as far as possible, which women are likely to need the highly technological aids available in hospitals. For the majority of women, however, pregnancy and labour are uneventful, and it is arguable whether any advantage is gained by giving birth in a hospital. Nevertheless, a belief in the greater safety of hospital deliveries (for the baby) is usually emphasised, and this is carried to excess by some doctors. Indeed, in one survey of women's experiences of childbirth in a London borough,[14] one woman said: 'The consultant said the best place for fathers was in the pub, and that home confinements were *almost like murder*.' Even before women reach the hospital doors, they have been taught to see their pregnancy as a medical (and pathological) event – a time of vulnerability and of suffering, when they need 'care' rather than assistance. Adrienne Rich, in her book *Of Woman Born*, discusses the significance of this suffering:

> Patriarchy has told the woman in labor that her suffering was purposive – was *the* purpose of her existence, that the new life she was bringing forth (especially if male) was of value and that her own value depended on bringing it forth . . . she has found herself at the centre of purposes, not hers, which she has *often incorporated and made into her own*. [our emphasis][15]

No one has yet been able to study the effects of home delivery adequately and to compare it with hospital deliveries. At present, the only women likely to have home deliveries by choice are going to be middle-class and articulate, and able to fight for what they perceive as their rights. The 'average' woman giving birth in a hospital may not be in this category. Because of these social differences, we cannot simply compare one with the other and say that one place is better than the other. Home confinements require the service of domiciliary midwives, but their training is being phased out under present health service plans.

The aim of this chapter is not to go into the arguments for home confinement very deeply – this has been done numbers of times, by lay people and medical professionals alike (see, for instance, Kitzinger and Davis[16]). Our aim is to look a little more closely at the consequences for women of the fact that birth now overwhelmingly takes place in hospitals.

GOING INTO HOSPITAL

In our labour wards today we have virtually no risk at all of such horrors as death from puerperal fever. The public health movement and the discovery of the role of germs in disease in the late nineteenth century have made the act of giving birth considerably safer. Nor indeed do we risk loss of our lives in order to save the child by Caesarian section. Nor yet do we have to suffer the agonies of difficult and lengthy labours, only to have it ended by the probably even greater agony of having the child removed in pieces. As Adrienne Rich points out:

> Repeatedly and for centuries, hooks were used to extract the fetus in pieces – a practice appropriately known as 'destructive obstetrics', with subdivisions including craniotomy, embryotomy, hook extraction and amputation of limbs. This was the speciality of the male physician as taught by Hippocrates and Galen;[17] Galen specifically declared it a male domain.[18]

Yet, despite these advances, medical science has contributed to the labour room in a way that may be less than desirable.

The hospital is a complex system for the management and care of sick and convalescent people. There have been two main effects of transporting birth into the hospital: first, the emphasis is on curative medicine, on pathological problems of illness, their discovery, measurement and cure; second, women, at an extremely vulnerable time of our lives, are put into a system that seems to have very little to do with us as individuals, but is concerned only to process the greatest number of women through birth without incident. In the process, the woman is placed in a position of compliance with expert advice throughout her pregnancy and delivery, and her personal needs and wishes tend to be largely trodden over. The hospital regime pivots around the ideas of illness, and of compliance by the 'patient' to expert advice: birth is increasingly seen as something that cannot be left alone, that must be interfered with, monitored, 'helped along'.[19] The mother is expected to fit into given images and concepts – she is expected to comply with doctors' instructions (even when not fully explained). Her will is often in opposition to the desires of the medical profession, but is not harnessed to make birth easier for all around – she is 'done unto' with a battery of machines and aids, and expected to submit quietly.

It has been suggested[20] that the 'exaggerated prudery' of Victorian times contributed to the situation that persists today, in which woman is passive, acted upon, even while giving birth. Control over labour now lies in the hands of a patriarchal medical profession. Women comprise a tiny percentage of gynaecologists and obstetricians in England and Wales. While midwives are nearly always women, it is still largely the male obstetrician who has the ultimate power over events. Furthermore, in talking to some women at whose birth a midwife was present, we have realised just how often those midwives perpetuate the same system of control over the labouring woman. In labour, woman is powerless: she has no control over her body. Adrienne Rich again:

> As long as birth – metaphorically or literally – remains an experience of passively handing over our minds and our bodies to male authority and technology, other kinds of social change can only minimally change our relationship to ourselves, to power, and to the world outside our bodies.[21]

There are many educational programmes on both radio and television about pregnancy and childbirth that contribute to a woman's image of what she *should* experience. These images are reinforced by doctors and clinics. The focus of such programmes is almost without exception on events of medical significance,[22] rather than on events that might be of great significance to the woman herself – such as feeling the baby move for the first time. Programmes typically portray the pregnancy as a series of visits to ante-natal clinics, with medical experts and high-powered technology taking a prominent place in the woman's life.

The emphasis is usually on the needs of the baby rather than the mother, and if not directly stated this theme is always implicit. One study of media images of pregnancy noted: 'This concern with the baby and its needs requires not only that mothers submit to medical procedures, but that they abrogate responsibility for major decisions concerning their pregnancy to their doctors.'[23] Interestingly, they go on to note that programmes about the post-natal period stress the responsibility of the parents, after the period of minimal responsibility implied by the programmes on the pre-natal period. 'Parents' is usually a euphemism for 'mothers', and 'mothers' are nearly always married, and devoid of any independent economic status.

The fact that the most advanced technology is found in the largest hospitals means that women find themselves entering a totally alien routine. Labouring women are forced into the mould of a routine designed for the care of sick people. Routines are not made to be disrupted, so birth must be 'fitted in'. As in other media images of medicine, to be a 'good' patient is to comply with the rules and the routine. The image of the passive, compliant, obedient patient does not tally with the birth process – unpredictable, often long and hard, *felt*, and where each woman will react very differently. The 'good' mother goes to ante-natal classes, follows all the textbook advice, doesn't create a fuss while in hospital, fits in happily, and emerges with a smiling, content baby and husband. In fact, everything has encouraged her to be distant from the situation, to hand over control to others, and to take advice – and she may be totally incapable or unwilling to take back control after the birth. She may suddenly be alone, with no real knowledge of her own feelings or her baby. The fact that delivery takes place in a hospital is only the end result of an entire process.

A stay in hospital may seem 'like a holiday' for some women, but, as Dana Breen points out, it is a holiday with complications:

> It is a holiday, not just in the sense of meals being prepared but in the sense that all worries and conflicts are temporarily suspended, that no 'working through' of difficulties takes place. This may seem pleasant at the time, but what about the long run? . . . It seems that what happens 'inside' the woman's mind was temporarily suspended in that week in hospital. Unfortunately, on returning home, many mothers are over- whelmed by anxiety, the anxiety which has been suppressed until then, with the detrimental effects this can imply for mother and baby.[24]

The fact that the woman's family cannot share in the experience of childbirth increases the tension within the family on her return home with the new child.

Many mothers, whether first time or not, are aware of a sense of competition between women in the ward: who has the biggest baby, the healthiest, which one is feeding fastest, who had the fastest labour. As long as the woman remains in hospital, she may be subject to this form of stress.

But perhaps the saddest part of this is that the medical staff often collude in it. Nurses, for example, may unwittingly create anxiety in the new mother by comparing her baby with others in the ward, or by judging her 'performance' as a mother against some 'norm'. At a time when the woman wants and needs to learn about her new infant, she is being taught that her child is not as good as, or is better than, the infant next door, or that she is 'failing' or 'succeeding' in some test. The atmosphere of competition isolates the woman in the ward. She is usually expected to dissociate herself from her body and its behaviour, to remain in control of herself and behave 'well'. The woman who screams in labour, or who cries afterwards, is often made to feel that she *ought not*, that she has lost control, that her own feelings are not natural, or that she should not give in to them. She is persuaded that to do so is to 'cause trouble', and is expected to acquiesce.

Unfortunately, the medical profession is not yet sufficiently aware that there is a problem. Most studies of women's attitudes after their deliveries indicate a favourable response – they do not often admit that the staff were in any way inefficient or unfeeling. This is probably because women do not like to risk offending the 'experts' in their own territory – because they are usually asked about it *while they are still in the hospital*. [25]

The British National Health Service is understaffed, and constantly facing economic difficulties. Nursing staff cannot be expected to stay with any one patient throughout her labour. The conservative medical profession, represented by the British Medical Association, considers that the only way to help the ailing NHS is to increase private practice: this is called 'increasing freedom of choice'. Freedom of choice, that is, for the minority who can afford it.

The contraction of health service expenditure has led to a move towards fewer and larger hospitals. The cottage hospital and GP units – in which a woman is attended by her own doctor rather than an unfamiliar obstetrician – are being phased out, as it is considered that they are less 'cost-effective'. [26] In other words, it is too much trouble and money to allow women, under the NHS, the option of having their own, familiar doctor in attendance, in a small-scale, local hospital.

TECHNOLOGY IN BIRTH

Technological intervention in the process of birth is becoming more and more common. In Britain approximately 10 per cent of all deliveries are instrumental – that is, they involve some intervention, such as forceps, induction or Caesarian delivery. Instrumental births can also earn the obstetrician more money. In the United States many more births are instrumental. Most people there have to pay for their medical care, with the result that medical practice is strongly influenced by the cost of different procedures. It has been pointed out that this may well be a factor in explaining why so many more surgical operations, such as hysterectomies, are carried out in the United States than here.[27]

Obviously, a certain number of interventions can be medically justified: Caesarian or forceps deliveries are usually necessary when a labouring woman is experiencing considerable difficulties. If these interventions did not take place, her life might well be in danger, as well as the life of her unborn child. Nevertheless, we would ask two questions: first, are all interventions absolutely necessary; and, second, are they carried out by the best means possible with the least discomfort to the mother, and with her full understanding of why and how the intervention is to take place?

One procedure that is carried out almost routinely in the majority of British hospitals is *episiotomy*. This is cutting the tissue between the vagina and the anus, as the baby's head is pushing through the cervix (the neck of the womb). It is done to prevent tearing of the vagina, which can, it is claimed, cause more damage to the tissues and can require more stitches. There has been considerable controversy over the incidence of episiotomies lately, as many women feel that they are done routinely even when they are not necessary. They argue that episiotomies are possibly more 'necessary' in our culture than they might be, owing to the use of the dorsal or lithotomy position in labour (that is, lying on your back for delivery). This position is used even more in the United States, where women often deliver on obstetric tables, flat on their backs, with their feet in stirrups. In Britain, however, many hospitals encourage a half-sitting-up position, supported by pillows. In most other cultures, and indeed in our own until this century, women gave birth sitting upright, over a 'birth stool', thus allowing gravity to help the birth. The opponents of routine episiotomies argue that in this position there is far less risk of tearing, and hence far less necessity for being cut. Needless to say, doctors are divided on whether birth positions influence the ease of a woman's labour.[28] It has also been pointed out that the lithotomy position greatly helps the doctors in their role of acting upon the 'patient', at the same time denying any active role to the woman herself.[29] In one survey, a large proportion of women who had stitches[30] found that it was a painful experience, especially following hours of effort in labour. Stitching may also leave knots of tissue that take months or years to dissolve – possibly even longer than torn tissue would have taken to heal. In view of this, why does the medical profession view the operation as so trivial that it can be done routinely, and often by a medical student who has had very little experience of deliveries? We have heard that in at least one medical school students were told that they *should* carry out episiotomies 'as a matter of routine'. Basically, 'it seems to be one of those things that doctors and midwives have become used to doing without enough thought as to whether the particular mother needs it, because it is part of the routine'.[31]

Induction of birth was originally used to start labour in women who were overdue, or to facilitate labour in women who were having difficulties. More recently, it has aroused considerable controversy in the pages of medical journals, following reports suggesting that induction was being used increasingly to arrange for births to take place during 'social hours'. While, in 1963, induction was carried out in only 13.75 per cent of births, by 1973 this had risen to 37.2 per cent,[32] and is still rising. In one study, for example, the researcher found that significantly fewer births took place in local hospitals on weekends or holidays![33]

Induction usually involves one of two types of hormone (oxytocin or a prostaglandin), which is fed into the woman's bloodstream. The hormone affects the womb, causing it to contract. The major risk of this procedure is of the dose being too high, causing the womb to contract very forcibly, which is painful for the woman and risky for the baby.[34] To prevent this, the drug is usually administered in the form of a drip, the dosage and timing being controlled and regulated by a machine.

Induction may become necessary when women are given painkillers – either such drugs as pethidine, or via epidural anaesthesia (anaesthetic injected into the base of the spine).

Where around 30 per cent of births are induced, a further 20–30 per cent of labours will be accelerated . . . available evidence suggests that [induction and epidural anaesthesia] may be associated with a raised incidence of Caesarian sections, forceps deliveries, and maternal infections. . . . The use of acceleration is now far more general than can be accounted for by selective application to cases of potentially long labour.[35]

In the end, techniques used to speed up labour and reduce pressure on staff may require more nursing and supervisory time and care.

Most painkillers, and especially epidurals, slow the progress of the labour, so that an induction agent is often considered necessary to speed things up again. One drug to slow us down, another to speed us up! Furthermore, epidurals effectively 'deaden' sensation, so that women have no control over their pushing, and forceps usually have to be used to deliver the baby.

During labour, a woman will often be surrounded by machines such as foetal monitors. These use either ultrasound or electrodes attached to the baby's head or the mother's abdomen, and they record the baby's heartbeat. This enables doctors to determine whether the baby is undergoing any undue stress as a result of the delivery. However, they do cause considerable discomfort and even pain to the mother, who is rarely informed of what is going on, or why the machinery is necessary. This entire experience can be, and is, very frightening to many women who do not understand the machines and are intimidated by the doctors so that they feel afraid to ask. With the heavy emphasis placed on the welfare of the baby (because, statistically speaking, the baby is more vulnerable), rather than the woman herself, there is a likelihood that a labouring woman feels coerced into accepting monitoring machines, even if she really would prefer to have an uninterrupted labour. This is one consequence of the 'over-medicalisation' of birth, which women often feel obliged to accept.

The rationale for most of the technology used in childbirth is that its use decreases the risk of death or injury to the baby. In support of this argument, statistics are often quoted showing that perinatal mortality (that is, stillbirths plus the number of deaths during the first week after birth) has fallen dramatically in the last thirty years. Doctors point out too that perinatal mortality has declined since the advent of chemical induction.[36] The implication is obvious – that hospitalisation and/or induction have somehow *caused* the lower mortality rate. However, this obscures all the other factors that might have contributed to the lower rate, factors that are not directly related to hospitals or to induction.[37] One factor that profoundly influences perinatal mortality is socioeconomic class: poorer women are less well fed, may be less well educated, have inadequate housing and so on. Richer women are better fed, can often take the time to rest and inform themselves if necessary, have adequate housing and access to private medical care if they wish. The state has tried, to some extent, to redress the imbalance.[38] We are all slightly 'richer' in the ante-natal care we receive on the National Health – iron tablets, vitamins, and routine monitoring of our pregnancies. We receive these benefits, however, at the price of signing over our bodies to the 'experts' and the machines.

Whether any woman accepts these procedures, or rejects them in the belief that it is better for her and her baby to go through labour without interference, should be entirely her decision. However, while it should obviously be a woman's decision what to do with her body, that decision must be based on adequate knowledge of the procedures involved. How can we decide what to do with our bodies if the information is withheld? Women are often resentful of not being told what the drip is for: many assume it to be simply a dextrose (sugar) drip. Often, the doctor says something like: 'I'm just going to put up a drip to speed you up a little,' without explaining anything.[39]

TECHNOLOGY AND THE MEDICAL PROFESSION

Perhaps the most worrying thing about the treatment of women in labour is not so much the presence of machinery, but the attitudes of those who have control over the new technology. These attitudes stem much more from the training and ethos of medicine than from any overt dislike of women. Scientific medicine requires a degree of passivity, of being acted upon, on the part of the 'patient'; and this is no less true in delivery. In the prudish climate of late Victorian Britain, doctors who practised obstetrics and gynaecology were extremely careful to emphasise that they did not see women as 'sexual beings' but as medical objects.[40] This divorce between women and their bodies is still apparent today. One doctor recently distinguished between the 'objective' and the 'subjective' in obstetrics, and pointed out that scientific medicine concentrates almost exclusively on the objective: foetal heart-rate, rate of contractions, or whatever.[41] The subjective experience of childbirth is of no concern or relevance to scientific obstetrics. An understanding of this subjective process plays little part in the training of doctors and midwives. Rather, they are invested with a knowledge of the 'science' and 'technology' of pregnancy and childbirth, and as holders of that knowledge are put in a position of authority over their 'patients'.

Making medical knowledge 'scientific' has resulted in a fundamental change in the nature of training. If we take the specific example of episiotomy, students are now trained to perform a routine operation, learning how to make a cut rather than learning to test the degree of elasticity of the vagina.

These 'scientifically determined' rights and wrongs may in fact be as transient as fashion. Recent years have seen the pendulum of medical fashion in full swing over the pros and cons of breastfeeding. Yet we women 'beneficiaries' of this knowledge are persuaded that it is 'objective fact'. There is all too frequently an attitude on the part of the professionals that 'we know best'. Women are expected to succumb to their authority, their power. Whether we discuss induction, epidurals, painkillers in general, episiotomies or anything else, the story is the same: the woman herself is often not told enough to enable her to make a real choice, or she is told that it is 'necessary' for her to accept the procedure. Women, it seems, should be seen but not heard.

Apart from the tendency to withhold information, a strand of misogyny is sometimes apparent. Many women have swapped stories about how misogynistic gynaecologists can be: one wonders why they choose a speciality dealing only with women! That there is almost a punitive element to these attitudes is illustrated by what many women *say* after the experience:

> 'You'd better put up with this, it's going to get a lot worse before it gets better.' [justifying the internal examinations that one woman was finding very painful, and was begging not to have][42]

> 'Your baby will be better off out than in.' [thereby justifying an induction][43]

> 'It's as if the baby's theirs. I actually got told off for holding my son!'[44]

Understaffing and overwork mean that even those doctors and midwives who are extremely sympathetic and careful to explain every procedure they are about to use find it difficult constantly to be aware of the specific needs of any one woman.

There have been efforts to promote 'natural childbirth'. 'Natural birth' simply means a delivery in which the mother accepts no drugs, or direct interference. She goes through the labour fully conscious, but relaxed, and more in control of the situation. If problems arise, then doctors may suggest intervention (for example, forceps delivery) if they consider it necessary. 'Natural birth' has enjoyed much enthusiastic support, especially as the mechanisation and dehumanisation of birth have increased so drastically. The supporters emphasise the pleasurable – even sexual – feelings that may result during the labour.[45] A birth that the mother can experience, without the use of drugs, might well be better for the child being born. It is also arguably better for the mother: our ability to produce babies, to go through the experience of giving birth, is unique to us as women; to give birth in a drug-haze is to be denied that experience.

However, the current trend towards advocating natural birth does have some disadvantages. There is a very real risk of some women being made to feel guilty, as a result of exposure to this publicity, if they do *not* have a completely intervention-free labour. Some women may feel guilty because they do *not* enjoy the experience, for whatever reason; because, for some of them, the pain becomes unbearable, and they have to request painkillers. The way in which a woman goes through labour should be *her* choice, free from unwarranted intervention by 'experts', but also free from other social pressures.

A major problem with a demand for a return to home confinements is that the technology that is necessary is being designed for use in hospitals and not for the home. To protect fully those women who, in a home confinement, suddenly develop complications would need a system of aids designed quite differently. As it now stands, our 'choice' is between a hospital birth with all its impersonality, and a home birth in familiar surroundings, but with some 'risk'. We do not consider this to be an acceptable choice.

THE SUBJECTIVE EXPERIENCE IN CHILDBIRTH

Whatever the dangers of childbirth centuries ago (and let us not forget that they were many), it was a woman's event. The labouring woman had a midwife in attendance, and possibly one or two other women. Their function was to help her through the ordeal, to understand, to sympathise. It is only in the last two centuries that male presence has been allowed in the labour room – with the advent of a male-dominated medical profession.

The experience of childbirth for women in this country is often extremely alienating. Women are told that they must go into hospital, 'for the good of the baby'; that they must submit – unquestioningly – to this or that medical procedure; that they may not be able to have their husband/lover/friend with them throughout the birth. To the woman who has never given birth before, the unknown is before her: terror is still present despite the advent of new drugs to kill the pain.

Some literature has been appearing which attacks the use of specific technologies surrounding childbirth, but less has been written about the more general effects on women of going into hospital to give birth. For many women, giving birth in hospital is alienating not just because of the onslaught of machines, but also because of the nature of the hospital environment. Many women may have to travel miles to their places of confinement – and so may their relatives, friends, and children. Visiting hours are often arranged for the convenience of the staff rather than for the visitors. In most hospitals women are allowed to have visitors only at specified times; between these times they are subject to a regimen that is far from the usual tempo of their lives, out of touch with their normal world. As a result of this alienation, 'Childbirth is often experienced by women as a situation in which they are being put to the test, and particularly in which their femininity is being put to the test.'[46] Our ability to bear children is often subtly denigrated in the treatment we receive and the attitudes we encounter. In the act of fulfilling our femininity by giving birth to a child, we seem to confirm our powerlessness.

In this acquiescence, this apparent passivity, we can detect an age-old acceptance of women's fate: in pain and travail do we bring forth, and this we should accept quietly.

> Passive suffering has thus been seen as a universal, 'natural' female destiny, carried into every sphere of our experience; and until we understand this fully, we will not have the self-knowledge to move from a centuries-old 'endurance' of suffering to a new active being.[47]

SOLUTIONS?

What are the possible solutions now open to us, and how should we be framing our demands for change?

The real solutions must come in a general restructuring of the care women receive throughout pregnancy, childbirth and its after-math.[48] Ante-natal classes must be made more relevant to the ex-periences of women. At present, as we discussed, they are geared to an explanation of medical events, and the technologies that labouring women may encounter. Perhaps if the classes offered women space to explore their subjective feelings about pregnancy, and to share their knowledge and experience with other women within a network of support, more women might attend them. For ante-natal classes *do* have an important part to play, especially in preventing or forewarning of possible complications. Women, especially poorer women, *need* the monitoring services, but while these are offered within an authoritarian and depersonalised setting, it is unlikely that all women will want to 'take advantage' of them.

If women are to give birth in hospitals, then the maternity units must become more sensitive to their needs. In particular, the people who care for pregnant women must be familiar to them. The Association of Radical Midwives points out that 'if it were possible for midwives to follow the same woman right through the process [childbirth], midwives would get more job satisfaction, and we think, women would have better care'.[49] The Association was set up in 1976 by midwives worried both about the effects on women and medical practitioners of the technological takeover of childbirth, and the phasing out of basic midwifery skills.

> It is upsetting and exhausting to have to stand powerlessly and see women deprived of the information and choices that are their right. In the labour ward, it sometimes becomes difficult for the midwife herself to remember that this woman is experiencing something unique in her life.[50]

In particular, they fear that phasing out district midwives, and demotion of hospital midwives to 'handmaidens to the obstetricians and their technology', means a loss of 'choice' for women. If we are to assert our right to make a choice about how and where we give birth, we must demand a restoration of the midwife's role and skills, particularly a return to domiciliary midwifery, and a reintroduction of GP units in hospitals, to provide a continuity of care.

We need to ask fundamental questions about the *kind* of health care we want. It is clear that improvements in the treatment of women in pregnancy and childbirth must not only be in terms of more and better services, but also requires completely rethinking women's health care. Women's decisions must rest on our conceptions of ourselves as women and mothers. We will not have better health care until we insist on our value as women, and on the subjective and personal value of childbirth.

It is time that we reclaimed our bodies, our childbirth. We must insist on giving birth in the way *we* want. We must insist on knowing *why* technology is required, if at all. We must insist on fair and human treatment during our labours, perhaps to be shared with someone whom we love (and who need *not* be a husband, as some hospitals insist). We must insist on a de-medicalisation of birth; it is an event full of mystery and importance to most women (and often to the men in their lives, if they have any), which profoundly affects our whole being, and it should not be treated as simply another medical event. We are more, much more, than the owners of uteri, cervices and vaginas. It is high time certain members of the medical profession came to recognise that.[51]

> To help women through childbirth is to share a mystery, and a miracle. . . . It is also to touch life at a key point in the social system, when generation gives way to generation, when bitterness and frustration and a sense of worthlessness can be handed down to the next generation, or the opportunity taken to grow in understanding and love. There is more to having a baby than simply pushing a [head] out into the world. Birth is also implicitly an assent to life. Those who help women in childbirth have the privilege of sharing that act of assent.[52]

'Breast-feeding gives mothers a fierce joy . . .'

Our thanks to Jalna Hanmer for her comments on an earlier draft of this chapter.

9

Contraception:
The Growth of a Technology

Vivien Walsh

INTRODUCTION

Millions of women welcomed the Pill when it was introduced in the United States in 1960. Within five years 26 per cent of married women (6.4 million) under the age of forty-five in the USA were current or past users, a further 19 per cent were considering it in the future and only 3 per cent said they had not heard of it.[1] Britain and other technologically advanced countries were quick to follow the American trend. Today about 50 million women throughout the world take the Pill, and in the West it has become virtually synonymous with birth control. Justified fears of the unwanted effects of the Pill were to come later. But in the 1960s it seemed that science and technology, in producing an effective and acceptable contraceptive method, had, through the agency of the drug industry, at last met an important and immediate need long felt by millions of women.

Historically, however, the Pill was a belated development in the technology of birth control. Most of the basic scientific and technical prerequisites for its development were in existence thirteen years before any company initiated any research or development work aimed at producing a birth control pill, and twenty-two years before it was first sold. The Pill was the first major innovation in the field of birth control for eighty years, although a significant section of the population had been practising family limitation during that time.

Contraception has been practised among small minorities from the earliest times, but effective family limitation among social groups of any significant size dates from the 1880s. In this period it was Britain that established the trends, as it was to be the United States in the postwar period.

The twenty years following the Bradlaugh–Besant[2] trial in 1877 were the years of development in the manufacturing and retailing of traditional contraceptives. 'These two decades saw the emergence of contraceptive techniques, of patterns of consumer preference and of retailing methods which were to survive over the next 80 years.'[3]

During these eighty years, old methods were improved partly as a result of innovation in other technologies: the vulcanisation of rubber, for example, revolutionised family planning as well as transport.[4] The latex process further improved rubber contraceptives, and some new chemicals were discovered for use as spermicides; for example, Volpar resulted from work by the Birth Control Investigation Committee, a body set up by the birth control movement to do research. But introduction of a distinctly new *method* of contraception had to await the development of the Pill. 'With the exception of the oral contraceptive,' writes Peel, 'there is not a single birth control method in existence today which was not already available, and available in greater variety, in 1880.'[5]

This is not a reflection of precocious developments in contraceptive technology in 1880. On the contrary, it reflects the fact that innovation had reached a dead end. Condoms, diaphragms, intra-uterine devices (IUDs) and spermicidal creams and pessaries were all in use by then, although they were not all equally reliable or safe (the IUD, for example, was not considered very satisfactory by several birth control teachers), and a great many trade names had been registered and patents granted for them.[6] Very little research was carried out over the next sixty years. The study of new materials for IUDs, for instance, developed *after* the advent of the Pill.

Why was so little research done? Why did it take so long for modern technology to be applied to the problem of birth control?

ATTITUDES TO BIRTH CONTROL

By 1938 it would have been scientifically possible to begin a project aimed at the development of a birth control drug. Progesterone and the other sex hormones had by then been isolated in pure form. Progesterone had been shown to prevent ovulation in certain female animals. A synthetic progestin (a compound that has an effect similar to progesterone), which unlike progesterone was active orally and at low doses, had been prepared. Two years later a cheap vegetable source of the compound from which the sex hormones could be manufactured was found to replace the expensive animal sources previously used.

The decision to look for an oral contraceptive was not, however, taken until 1950. Until then, there was no interaction between the chemical, biological and clinical studies that might have been relevant to a birth control pill: they were carried out in different organisations and mostly without contact between the researchers. Some of the work was done, at least ostensibly, for other purposes. For example, clinical studies of sex hormones had been made in a search to overcome *infertility*. Drug companies had developed drugs containing sex hormones and related compounds to treat gynaecological disorders such as painful or irregular periods.

The war was partly responsible for the delay. Some researchers were working on other projects for the war effort, while many were in the forces. Birth control was not considered important to the war effort (see the section below on policies towards population and birth control). Indeed, the supply even of traditional contraceptives was imperilled by the acute shortage of rubber.[7]

The war, however, was not the only reason for the delay. Many scientists involved in basic research in the science of reproduction, as well as those concerned specifically with birth control, have testified that their work was impeded by age-old and deep-rooted prejudices and social taboos. 'Religious and social prejudices', writes Parkes, a leading physiologist in the field of reproduction, 'have caused matters relating to reproduction . . . to languish under a taboo strangely at variance with private practice.'[8] 'The whole area of reproduction, let alone contraception,' confirmed another historian of the Pill's development, 'was not entirely socially acceptable.'[9] Several scientists who had worked on problems concerned with reproduction have all noted this prejudice.[10] Researchers working on birth control as such even risked being thrown out of their laboratories: 'For Baker,' wrote Peel, 'contraceptive research . . . is permanently symbolised in his recollection of assembling his apparatus and reagents on a handcart and trundling this from department to department.'[11]

The medical profession was, for years, strongly opposed to birth control, arguing against it on both moral and medical grounds. At the end of the nineteenth century, when the practice of birth control was first beginning to spread, doctors – despite, as will be shown, their own practice – claimed that its effects 'included galloping cancer, sterility and nymphomania in women; mental decay, amnesia and cardiac palpitations in men; in both sexes the practice was likely to produce mania leading to suicide'.[12] In the 1920s contraception was still considered 'distinctly dangerous to health', sterility and 'mental degeneration in subsequent offspring'[13] being among the alleged effects.

Doctors are obviously in a key position to help promote or inhibit innovation in drugs or in birth control techniques because they prescribe medicines and fit or recommend other medical and related supplies. Their prestige and influence can also be used to counteract prejudices or – more usually the case historically – contribute to reinforcing them. In this case, the medical profession took no initiative in getting the provision of birth control advice extended,[14] and, far from doing clinical or scientific research themselves,[15] most doctors were strongly opposed, in general, to studies being carried out by themselves *or* anyone else. Between the wars medical students were prevented from inviting birth control clinic staff to give lectures to their societies,[16] and as recently as 1957 only twelve out of thirty-six UK medical schools gave a formal course on contraception, five without clinical instruction.[17]

Doctors, in fact, were among the earliest to use birth control: in the late nineteenth century they appear to have practised family limitation more extensively than any other social group.[18] But publicly, as a profession, they nevertheless continued to maintain a resolute opposition to birth control. Their opposition reflected two interrelated objections. Birth control conflicted with the Victorian social mores which they devoutly upheld; and contraception was unmistakably associated with publishers, practitioners and tradesmen who were far from being considered respectable. No matter that it was the medical profession's aversion that consigned contraception to its own clandestine milieu; 'doctors had a reflex aversion to anything that smacked of lay medicine, sensationalism or quackery'.[19] Until well after the Second World War really acceptable techniques for birth control, reliable ways of assessing the effectiveness of different methods and clinical studies were all lacking, so there was inevitably some justification for speaking of birth control and quackery in one breath.[20]

In the interwar period, 'the demand for contraceptives outstripped the supply of medical knowledge and gave rise to a large industry riddled with quackery and dishonesty'.[21] Would this have been the case if the medical profession had been prepared to give advice and prescribe supplies? If they had carried out clinical studies and cooperated with those in the birth control movement who had done research? If they had encouraged the major pharmaceutical firms to enter the field?

Doctors' opposition was not confined to their refusal to accept contraception as a facet of medical treatment. They were all the more hostile to birth control as a technique that might be used for social, economic and other non-medical reasons. All but a small minority of the medical profession have always held an extremely narrow view of medicine in general: few doctors concerned themselves with preventive medicine or health care in its widest social sense. And most doctors instinctively felt that involvement with techniques that stretched across the traditional boundaries of medicine represented a dire threat both to their learned authority and to their professional status.

When the profession at last began to move away from total opposition to birth control, doctors then adopted the view that contraceptive advice should be given only to women whose life or health was in danger. As with abortion later, this meant in effect a change

'. . . an experience of passively handing over our minds and our bodies to male authority and technology . . .'

(from an original cartoon by Pen Dalton. Reproduced with permission.)

from outright hostility to an attempt to take complete control, since it would be doctors themselves who would judge whether life or health were in danger.

Birth control, of course, was not the first case where the medical profession had been suspicious of innovations from outside the profession or 'lay medicine'. Pasteur, who was a biochemist, chemist, physicist and crystallographer, but was not a doctor of medicine, had to battle for years against that attitude. [22]

Even the retailing of contraceptives was separated from that of medical supplies. An article by J. Peel [23] describes the alienation of the former. Medical supplies were sold in chemists' shops, whereas barbers, surgical stores, mail order firms and, in the United States, tobacconists and petrol stations remained the main outlet for contraceptives for many years. As recently as the 1950s, letters to the *Pharmaceutical Journal* showed that many chemists were strenuously opposed to selling, advertising and displaying of contraceptives although they were highly profitable. Some did not even sell them until a few years ago because they were associated with 'promiscuity, vice and prostitution'. For many years, even where it was not actually illegal, the trade in contraceptives remained clandestine and underground in nature, and the numbers of people using various appliances could only be guessed at.

GOVERNMENT POLICIES TOWARDS POPULATION AND BIRTH CONTROL

Opposition to birth control

Until recently most countries, or at least those that favoured a growth in the birth rate, had pro-natalist laws and policies. The power and prosperity of states has historically been associated by their rulers with large numbers, especially of workers, potential consumers and, when necessary, potential members of the armed forces. Governments have attempted to encourage population growth through, for example, welfare benefits, tax allowances and a minimum age for marriage, as well as prohibition of birth control. [24]

In Britain, a government publication (1954) stated that 'the state played practically no part in the great spread during the twentieth century of the knowledge and use of birth control methods'. [25] In America, 'open association with contraception and its positive promotion was an activity which our government shunned like the plague',[26] and in fact 'all attempts to frustrate the process of parenthood have been discouraged by government and the courts'.[27]

In the United States, where the Pill was first developed and marketed, the Federal Comstock Laws (named after the senator who introduced them in the 1870s) and various individual state laws banned the sale, display, advertising and even the use of contraceptives. Contradictory regulations were adopted by several states. For example, Nebraska forbade the sale of contraceptives but set quality standards for those sold; Mississippi both prohibited contraceptives and offered birth control supplies through the state's own health services. (See W. Petersen, *The Politics of Population*, 1964.) Many of the research programmes financed by the American birth control movement in the 1930s had to be undertaken in Britain as a result of the legal situation. During this period, Margaret Sanger and her colleagues were continuously involved in legal battles. Although they managed to obtain several favourable court decisions, the continued existence of the laws provided an excuse which was used by the medical profession and the government to avoid taking any action about birth control. After the Second World War the legislation was less stringently applied. Pincus and his co-workers carried out their research on the Pill in Massachusetts, where anti-birth control legislation was not amended until 1966. But the laws were still activated from time to time to harass those who ran birth control clinics.

In the UK there was never direct legal prohibition of birth control, although laws concerning pornography (e.g. the Obscene Publications Act, 1857), for example, had from time to time up to the Second World War been interpreted to include contraceptive information. Neither the state nor the medical profession was, however, anxious to promote or sponsor birth control. In 1930 a 'permissive' memorandum (i.e., not circulated, but supplied on request to local authorities) was issued,[28] which was the outcome of years of pressure by the birth control movement. This gave local authorities the power, if they wished, to provide contraceptive advice to married women in their maternity and child welfare clinics, 'in cases where a further pregnancy would be detrimental to health'. Thirty-six authorities did take advantage of those provisions, but such was the lack of knowledge among their medical staff that they were often obliged to refer their clients to the voluntary clinics.[29]

By 1949 the Royal Commission on Population was recommending[30] that advice on contraception should be available 'to all married persons who wanted it', and that it should be a 'duty of the

National Health Service'. The recommendation was not taken up either by the government or the medical profession (although GPs began to prescribe the Pill when it was introduced), and in the 1960s the Family Planning Association (FPA) was still calling for birth control advice to be provided by the NHS.[31] It was 1974 before free contraceptive supplies and advice were available and unmarried women were officially accepted as clients by all the clinics.[32] During 1976, 446 FPA clinics came under the administration of the health authorities.[33] Recent cutbacks in public expenditure, however, have already begun to threaten this service.

British and American policies seem to have been a reflection more of a desire to 'safeguard' public morals than to increase population size; in other countries, restrictions on contraception were more explicitly part of population policy. France introduced the 'Code de la Famille' in 1940 in response to an absolute decline in population during the depression, by which marriage and childrearing were encouraged by various welfare and tax benefits, while contraception and abortion were restricted.[34]

The most extreme example of state intervention on population matters took place under fascism in Germany. The racially (and socially) 'inferior' were sterilised and/or sent to the gas chambers. Aryan women were prevented from working or from using contraceptives (as well as being denied many other rights) and subjected to intensive propaganda in order to force them to breed as rapidly as possible.[35]

On the other hand, Sweden has adopted a generally pro-natalist policy, which has taken the form of encouraging wanted children, by providing welfare benefits, nurseries and paid maternity leave, while making birth control freely available to those who want to enjoy sex but not to have children.[36]

Encouragement of birth control

What was it that stimulated the rapid change in prevailing attitudes, particularly official attitudes, at the time the Pill was becoming a practical possibility? It seems to have been above all a fear of a world 'population explosion'. This fear acted as the vital catalyst in the complex but quite sudden change of public, medical and government attitudes to birth control.

Sudden alarm at world population growth, however, was a political response rather than the reasoned analysis of demographic trends and their underlying social causes. A prevailing fear of overpopulation in the 1950s seems to have replaced a recent enough fear of underpopulation (in the 1930s) among demographers and economists just as rapidly and emotively as among politicians and journalists.[37] The doubtful validity of new 'consensus' opinion about the population explosion does not alter the fact that the strength of prevailing attitudes had a marked influence on government policy – and the plans of the pharmaceutical companies. (Company policy was also influenced by financial incentives – see the section below about the drug industry.) For example, population growth is widely believed to be the cause of starvation in the Third World, although it has been shown (e.g. in the *Economist*, 20 December 1969) that food production is now increasing at one and three-quarters times the rate of the population. Starvation is a social, not a technical, problem, requiring redistribution of resources; but it is population control that is posed as the solution in official policy.

Demands from political leaders in the West for measures to limit reproduction in the poverty-stricken countries of the Third World made it impossible to sustain generalised moral or social objections to contraception at home. Sooner or later, under the pressure of social and economic changes (described in a later section), contraception would have become acceptable. But the widespread fear of the 'population bomb' suddenly advanced its legitimisation.

As experience was soon to show in 1960, the better off, well educated, mainly white, women in the United States and the West generally were the drug industry's prime market for an oral contraceptive. It did not matter, given the new climate of opinion, that concern about rapid population growth in the underdeveloped countries had little to do with the desire of the more affluent social strata of the West to plan their children to suit their new life-styles. Most of the drug companies and scientists who have commented on the motivation of their development work cite the 'population explosion' as the main impulse.

In governmental and public anxiety, the scientists had found an acceptable reason for doing their research, and the drug companies for investing in a potentially profitable innovation. Gregory Pincus,[38] who played a key role in the development of the Pill, cites it as one of the reasons he began research in 1950. Marrian and Nelson,[39] who were both working on the biochemistry of reproduction, state that this gave an important impetus to research. *Science Newsletter*,[40] reporting early findings by Pincus and others, used the sensational headline: 'the explosive increase in world population could be squelched by a tiny pill'. This is not at all untypical of the popular and semi-technical reporting of that period.

Governments themselves, however, were still relatively slow to respond to demands for action on the population question, though they themselves contributed to the general alarm.

Shortly after the Second World War, proposals that the US Government adopt a definite population policy at home – or get the United Nations to adopt such a policy abroad – had very little support.[41] It was in the second half of the 1950s, the time when the Pill was going through clinical tests before being marketed, that policies began to change. In 1959 the recommendations of the controversial Draper Report – that the United States should support UN schemes for birth control in underdeveloped countries – were accepted.[42] In that same year, however, Eisenhower said of birth control: 'I cannot imagine anything more emphatically a subject that is not a proper political or governmental activity, or function or responsibility.' Six years later Eisenhower and Truman became co-chairmen of the honorary sponsors of Planned Parenthood – World Population; and Johnson became the first president to report federal support for population control.[43]

Public attitudes and official thinking were at last brought more or less into line with the widespread and fairly long-established practice of contraception.

Prejudice, taboo, policies and laws had retarded innovation in birth control and the spread of such techniques that did exist. Fear of the 'population explosion' and the consequent change in laws and policies in turn served to promote or speed up the use of birth control and innovation in technique.

SOCIAL CHANGES

At the individual level women, or couples, decide whether or not to have children and when to have them for a whole variety of personal reasons. At the level of society, however, general trends in fertility can nevertheless be observed which are related in a complex way to various social and economic factors. It is generally agreed[44] that fertility (defined by demographers as actual fertility, whereas fecundity is capacity for reproduction) is related to living standards: the wealthier a country, the more widespread is fertility control, it being most widespread in the wealthier sections of those countries.

The history of the spread of birth control is essentially the history of the social conditions that gave rise to the personal motivation for family limitation and at the same time created the means to make it possible.

Family limitation has been practised for centuries. Abortion, infanticide and withdrawal were used in the earliest societies, and even primitive barrier or spermicide contraceptive techniques have been reported.[45] But until the late eighteenth century or later, contraception was confined to small élite groups.[46] In most early, predominantly agrarian, societies there was no need to limit births. A multitude of children was needed to lighten the burden of work and even literally to ensure the continuation of the family, community or tribe. Famine, epidemics and wars were a cruelly effective means of preventing overpopulation.

On the whole, it is when living standards are high enough to offer a promise of future security that people try and maintain or improve their position by family limitation.

Agrarian societies that became more prosperous and stable tended to achieve a certain equilibrium in their populations. Demographic studies of eighteenth- and nineteenth-century Europe, for instance, clearly reveal a broad correlation between growth trends, family size and patterns of land tenure, the latter being an indication of the prosperity of different social groups.[47] France, for example, following the land reform established by the 1789 revolution, achieved a remarkable, prolonged stability of population.[48] This stability of total population and family size was achieved almost entirely by 'pre-industrial' methods of family limitation – abstinence, withdrawal and abortion.

In Britain the practice of birth control was followed in the late nineteenth century by upper-middle-class, business and professional groups. The aim was to postpone births until the education of the male 'head' of the family was complete and the beginnings of financial success achieved. Then tighter economic circumstances at the end of the century made them strive to keep up the consumption habits and standards of living to which they had become accustomed during the long period of preceding prosperity.[49] As with France's prosperous farmers, the Victorian 'paterfamilias' achieved his desired family size largely by non-appliance methods; but as patent contraceptives (mentioned in the introduction) began to become available they were increasingly used.

Historically, there is ample evidence to show that where – under certain conditions – there were good reasons for family limitation this would be achieved in spite of the limited methods available. The necessary conditions, of course, include social and economic circumstances that make it worth while to limit families. They also include a related level of culture and personal consciousness high enough to allow individuals the necessary degree of control over their own lives.

In poor communities, on the other hand, fertility is not limited voluntarily just because contraception is provided, even if it is free. The collapse of rotten social structures under the pressure of poverty, hunger and disease both destroys any stable social norms conducive to successful family limitation and denies individuals the self-consciousness and self-confidence that is clearly essential to the consistent practice of birth control.[50] Social and economic change in underdeveloped countries, involving an end to exploitation and a redistribution of wealth, would increase the living standards of the poor, so that even a fairly rapidly growing population could be adequately fed, clothed and housed. It would also lead to demographic changes that all the West's birth control programmes have failed to achieve.

In the West, the postwar increase in standards of living influenced attitudes to birth control. As more and more people found themselves in a position to enjoy improved standards of living, so more and more people began to appreciate the advantages of smaller families, and more flexible attitudes developed in which family limitation methods could spread. In 1959, immediately before the Pill was first introduced, 81 per cent of couples between eighteen and thirty-nine surveyed by Whelpton *et al.*[51] had practised family limitation.

Parallel with the fact that increasing numbers of families – which previously meant primarily the male 'heads' of the families – were feeling the need to plan their children and were enjoying the conditions that enabled them to do so, there was another vital development: a growing demand by women for their own extended education and their own careers.

Immediately after the war there was a renewed emphasis on the family. Many women who had worked during the war were obliged to become housewives again, especially in Britain. But by the 1950s women began to regain their wartime advances and went on to achieve more in terms of independence, education, careers, and freedom from their traditional role in the family.[52] In the postwar period, both in the United States and in Europe, there was an increase in the number of women, including mothers of children under six, who worked outside the home. The biggest increase was among married women and mothers.[53] According to Banks,[54] this was the most significant factor in the spread of birth control. (Figures for America are given in this section as it was there that the Pill was developed, and there that changing attitudes were most crucial to its being marketed.)

Chafe[55] shows that twice as many American women worked in 1960 as in 1940. The proportion of married women at work increased from 15 to 30 per cent; there were more than 10 million in 1960. The number of mothers in paid employment increased over four-fold, from 1.5 million to 6.6. million. In 1960 39 per cent of mothers with children aged six to seventeen worked outside the home, and they accounted for almost one-third of women workers. Says Chafe: 'Many couples found it impossible to fulfil their quest for a higher standard of living on one income alone.' The highest growth in the female labour force took place among well educated wives from families with already moderate incomes – the group most likely to practise, or want to practise, family limitation. Before the war, almost all women who had jobs outside the home came from working-class families.

Improved living standards were not the only reason for more women working. Almost 90 per cent of women interviewed (sample sizes not given) in one survey[56] said they liked their jobs, in addition to liking the independence of a pay cheque, particularly as they had the opportunity to be with other people and receive recognition for their work. This was true of manual as well as white-collar workers. 'I'm strong and I do a good job,' said one cafeteria worker.

'Two thirds of married women interviewed referred to their jobs as a basis for feeling "important" or "useful", while only one third cited the socially sanctioned activity of housekeeping.'[57] (Sample sizes not given.)

The number of women following further and higher education courses also increased. The number completing four or more years of college education increased between 1940 and 1945, increased more steeply between 1950 and 1956, and even more steeply after that.[58] Studies by the United Nations[59] and by Whelpton et al.[60] indicated that there is a positive correlation between the education and employment of women and the use of contraception. Thus, in general, the higher the educational level, the more likely a woman is to want to plan her children and also to take up a job or a career which requires the planning of her children. The greatest use of the Pill (in a study conducted in 1965[61]) occurred among young women and women who had had a college education. And, of course, a large number of unmarried women began to take the Pill, and in many cases opted not to marry at all. The emphasis on married couples in the trends mentioned here is a reflection of the statistical data available, which has tended to concentrate on married women as users of contraception.

THE BIRTH CONTROL MOVEMENT

The practice of birth control spread primarily under the pressure of a combination of social and economic changes. But such changes do not mechanically determine developments. Their influence is felt through the consciousness and behaviour of individuals and social groups. Historically, however, new techniques have not always immediately been translated into social practice. This was the case with birth control. For many years the contraceptives available – inadequate though they were – were used by only a small minority, although family limitation (by means, for example, of withdrawal or abstinence) was practised by many more people. Not only was innovation in contraceptive methods retarded: the diffusion of existing methods was also very slow. This gulf between the practice of family limitation and the privately felt need for contraceptives, on the one hand, and the lack of reliable birth control advice or publicly available contraceptives on the other provided a basis for the birth control movement.

The birth control movement tried to rescue contraceptives from the back-street retailers, attempting to evaluate the different techniques and to improve the best available. But their main battle was to try and break down the complex of moral, religious and cultural values and prejudices which denied knowledge of birth control techniques to the majority and confined contraception to its own clandestine milieu beneath the surface of 'respectable' society. The birth control movement, moreover, was animated by a dual motivation, and the ambiguity between the two elements was to prove problematic. It clearly championed the cause of women who wanted to choose whether to have children and when to have them. But it also raised the fear of over-population among the poor and needy (neo-Malthusianism).

Pressure-group activity on the issue of birth control dated from the Bradlaugh–Besant trial (1877). The Malthusian League was established in that year. Its members were influenced by the thinking of Malthus,[62] who held that the main cause of poverty was over-population among the poor.[63] While Malthus urged 'moral restraint' to limit population growth, the Malthusian League and other 'neo-Malthusians' advocated the widespread use of birth control methods. [64]

The League appealed almost exclusively to those 'who desire . . . to relieve the increasing burden of the middle class'[65] to support the spread of birth control methods in order to keep the poor from breeding. At the same time there were those who supported birth control as a means of improving the lot of women (although very few women were involved in the Malthusian League), who 'are kept down by their two-fold chains of economic dependence and enforced maternity'.[66] It was recognised that the 'demand for birth control represented in effect an attack on the . . . basic moral assumptions of a male dominated society: to challenge this was to undermine a primary assumption of the social order'.[67] This conflict has been evident among supporters of birth control ever since.[68] There were, and are, those who believed that women should be able to have children only if and when they want them. More and more women, and men, are choosing to live outside traditional family and social structures, and have seen the availability of contraceptives as one of the prerequisites for making a choice of life-style a real possibility.

Even in the 1920s Marie Stopes and her colleagues were putting forward ideas – inchoate though they may have been – about personal development and happiness as their principal reasons for promoting birth control clinics. Meanwhile, the Workers' Birth Control Group was formed, with Dora Russell as its secretary, in order to link the campaign for birth control and socialism. This was aimed chiefly at the Labour Party, and even though the labour leaders were not at all enthusiastic (regarding birth control as an issue likely to lose them votes) there was considerable pressure from within the party on this issue. The Labour Women's Conference in particular came out in 1924 in favour of birth control clinics, while Labour-controlled boroughs such as Battersea and Stepney set up birth control clinics in the same year – and were threatened with the withdrawal of their grant by the Minister of Health.[69]

On the other hand, there were, and still are, those who believed that birth control was the means of selectively preventing 'inferior' races or classes from breeding too rapidly. (This was the aim of the eugenics movement.) There seems to have been a definite element of this in the fear of the 'population explosion' put forward by drug companies in the 1950s and 1960s as their reason for developing the Pill; and by governments for legalising, and later sponsoring, birth control.

When birth control was first discussed in the US Congress in 1934,[70] for example, it was clear that some members would be prepared to support a birth control programme aimed at poor blacks in the south, while they would not support the availability of birth control for the population as a whole. The newspaper headline 'Bumper Baby Crop Threat to US' stated the attitude of the establishment baldly.[71] Meanwhile, much more recently, while too carefully formulated to suggest that they thought the poor were a threat to the rich, comments from Lyndon B. Johnson and Robert McNamara (president of the World Bank) nevertheless indicated that their concern was not exactly with providing the people of the Third World with the means to have the number of children they wanted when they wanted them: 'Let us act on the fact that less than five dollars invested in population control is worth 100 dollars invested in economic growth'[72] '. . . We are conscious of the fact that successful programs of this kind will yield very high economic returns.'[73] And: 'The ultimate triumphs of foreign aid are victories of prevention. They are shots that did not sound, the blood that did not spill, the treasure that did not have to be spent to stamp out spreading flames of violence.'[74] High population growth is thus advanced as a reason for the failure of economic development, rather than exploitation by landlords and capitalists at home and multinational companies from abroad. The unequal distribution of wealth and power, both between and within countries, is then posed as a biological problem with birth control programmes the necessary technical 'fix'.[75]

Those in the birth control movement between the wars who argued that spending on birth control could be a substitute for much needed spending in the fields of poverty, housing, health, etc., alienated many in the labour movement and other potential supporters. Some were opportunists: they believed that arguments about the relief of poverty were more likely to win them support in official circles than arguments about the oppression of women. Thus Margaret Sanger, who professed to be a socialist, and who at first had argued that birth control was a means of freeing women from the burden of too many children, later began to make statements such as: 'making birth control available to the poor would alleviate misery, ease unemployment and reduce taxes'.[76]

Some in the birth control movement, however, were openly associated with the eugenics movement (defined on p. 196). C. P. Blacker, for example, was secretary of both the Eugenics Society and the Birth Control Investigation Committee.[77] This confirmed the worst fears of those socialists and trade unionists who were opposed to birth control for political (rather than religious) reasons. There was thus, in the labour movement, strong support for birth control as a means of improving the conditions of women and of the working class generally. At the same time it was also strongly opposed, as, at best, a diversion from the struggle for social change and, at worst, a technique to be used for outright oppression.

Similarly, the issues of resources and the environment have been used more recently in both a socialist and a reactionary way. The depletion of resources and pollution of the environment have been used to argue for an end to the capitalist economic system that wastes and pollutes and is responsible for the grossly unequal distribution of wealth and production; but they also have been used to argue that the survival of the world demands a limit to economic and population growth.

In practice, no distinction is made between growth in standards of living of the poor and growth in wasteful production;[78] while 'zero population growth' – when advanced by well intentioned liberals just as much as, say, the South African Government in its development of contraceptive services for the black population – is usually proposed as an alternative to economic and social development.[79] Consequently, the labour movement has treated ecological issues in an ambivalent way, just as it regarded birth control in the interwar period.

In the long run, however, social and economic changes provided a suitable climate for the development of birth control techniques, at least in the West; and these changes were considerably enhanced by the activities of the 'birth control movement'.[80]

Those individuals and organisations described as the birth control movement not only campaigned, especially between the wars (fighting many battles in court), for the state to provide birth control clinics – and, where necessary, to change the legislation that made the use and provision of contraceptives illegal; they also organised most of the clinics in the UK, the United States and other countries, which made supplies and information available to women. They were responsible for propaganda about birth control among the general public; they put pressure on the medical profession to take some responsibility for contraception, and for many years (even after the Second World War) they were the only source of accurate scientific information about birth control and contraceptives.

In the interwar period, different tactics were adopted in different countries. In Britain, the Malthusian League came to the conclusion that the opposition of the medical profession would prevent the advance of contraceptive theory and practice unless birth control clinics were set up outside the medical establishment.[81] In the United States, where birth control was formally illegal, Margaret Sanger and the birth control movement deliberately set out to co-operate with the more enlightened section of the profession, having decided that medical support would be necessary to overcome the extra obstacle of illegality.[82]

In addition to the activities already mentioned, the birth control movement initiated and financed research. The Birth Control Investigation Committee (BCIC), established in 1927, grew out of the voluntary clinic movement and was responsible for most of the research in Britain, which took place on the periphery of established medical research.[83] Some US-financed research was also carried out in Britain.[84] Weisner's work, the first recorded study of hormones in relation to birth control (1928),[85] received finance from both Margaret Sanger in the United States and the BCIC in Britain. Pincus, one of those who in 1950 in the United States began the research on hormonal contraception that led to the Pill, cites a visit from Sanger as being a major influence in his decision to begin the work. She made funds available to him, although he also received a grant from the drug company, G. D. Searle, to whom he was already a consultant.

Today, organisations like the Family Planning Association are highly respectable sections of the community, having close links with the National Health Service and the local authorities. In the United States, leading public figures associate themselves with planned parenthood. But at the time of the early campaigns, advocates for the acceptance of contraception included groups and individuals of whom many were feminists, radicals and socialists, and their activities were not seen as being at all 'respectable'.

The birth control movement, then, was analogous to today's pro-abortion campaign,[86] and was frequently described by the establishment in the same terms. The medical journals, for example, were closed to contributions from Margaret Sanger and her colleague Dr Hannah Stone, describing them as 'sensational contributions from fanatical propagandists or hysterical ladies'.[87]

THE DRUG INDUSTRY

The 'Pill' transformed contraception. Not only were steroid contraceptive drugs the first major contraceptive innovation for eighty years, but their appearance itself made an important contribution to public acceptance of contraception. Associated with all the prestige of 'Science', and consequently soon winning the support of an increasing number of doctors, the Pill 'contributed not merely to the technology of contraception, but to the freer public discussion of birth control generally, which is now set within a professional medical context'.[88]

Yet the pharmaceutical manufacturers were slow and at first reluctant to associate their names with contraception. Their caution in this field was in marked contrast to their usual eagerness to promote innovations for which they see a profitable market. Usually, the drug companies are prepared to invest huge sums in advertising campaigns to persuade the public or prescribing doctors that a new pharmaceutical product is 'needed', whether or not such a need had previously been perceived. Indeed, their readiness to promote the use of drugs that may have serious 'side'-effects, or are intended for use as a technical 'fix' for problems with wider social and economic causes and consequences, has often been criticised (see next section). In the case of the Pill, however, for which there was theoretically a market among millions of healthy women who wanted to avoid or postpone having children, the companies were slow to move.

The evidence available at the time unmistakably showed that more and more people were practising birth control, while the inadequacy of available contraceptive methods suggested the need for a more effective and acceptable method.

An important survey carried out for the Royal Commission on Population by Lewis-Faning[89] showed that there was still heavy reliance on non-appliance methods after the Second World War, especially among the working class. In America there was probably slightly greater use of the sheath and the diaphragm.[90] The first nationwide survey of the extent of family limitation in the United States in 1954 showed that 70 per cent of couples between eighteen and thirty-nine had practised family limitation.[91] Nevertheless, a significant number of them, the survey also showed, had an average of two children more than they really wanted. Five years later, immediately before the Pill became available, a second survey showed that the percentage using contraception at some time had risen by 11 per cent to 81 per cent, the proportions using the different methods remaining about the same.[92] Decline in fertility and in family size over the period 1850–1960 is discussed in detail by Wrigley.[93]

However, while the market was growing, and business was

certainly very profitable, the traditional contraceptive manufacturers were not prepared to meet the need for an innovation in birth control methods. It was estimated that in the United States expenditure on contraceptive advertising, which was, strictly speaking, illegal, amounted to $935,000 even in 1932,[94] while American women spent more than $210 million annually on contraceptives.[95] But the industry was satisfied with its semi-clandestine, lucrative market, and was not interested in spending money on research, clinical testing, quality control or even market evaluation. The firms were certainly indifferent to the accuracy of their advertising copy. C. I. B. Voge, one of the few researchers who had made a serious technical study of contraceptives (*The Chemistry and Physics of Contraceptives*, 1933), and whose work had been sponsored by the birth control movement, criticised the manufacturer's complacency. They 'appear to be entirely unconcerned about their previous failures and they are emphatic that the new formula will be much more effective!' he said.[96] (On this occasion he was referring to spermicides.)

This attitude persisted to the 1950s, although lack of *clinical* research in particular was as much the result of hostility, ignorance or indifference on the part of the doctors as lack of concern by manufacturers.

Despite the potential market for a new and more effective method, the traditional contraceptive manufacturers were incapable of innovation. It would have meant reorientation towards research. It took the drug industry's entry, from outside, to the field of contraception to initiate significant technical progress. It was practically the first time that modern science and technology had been applied systematically to contraception. Yet despite the vast potential market, of which the companies almost certainly were aware, and the technical possibility of producing a pill, even the drug companies hesitated through fear of hostile public reaction to their involvement with birth control. 'The attitudes prevailing in 1959 . . . were vastly different from those which exist today,' said Winter of G. D. Searle:[97]

No major pharmaceutical manufacturer had dared to put its name on a 'contraceptive'.[98] The individual reaction of a very large religious minority in the United States could not be gauged. The possibility of losing overnight one fourth of all our personnel, a considerable portion of our hospital business and a crippling number of the physician prescribers of our products was not to be dismissed lightly.

Irwin, also of G. D. Searle, said 'We were a company with an absolutely impeccable reputation', whereas 'birth control was still a delicate subject. The word contraceptive suggested rubber goods from a back street shop, not a tablet from a leading pharmaceutical company.'[99]

G. D. Searle, in retrospect, likes to emphasise its pioneering role in first marketing the Pill. Certainly, the reasons for not taking the risk were strong enough to prevent Parke Davis from being first in the market. Syntex was the company that first synthesised and patented (in 1954) a suitable progestin, but at that time they had no marketing organisation to launch the final product. They offered to enter a licensing agreement with Parke Davis, who refused.

But despite the reasons for caution, Searle was also undoubtedly aware of the potential market. They 'had in their laps . . . the classic golden egg of the drug business . . . with a virtually guaranteed sale to millions of women throughout their child-bearing lives . . . a drug to be taken not by the sick but by the healthy'.[100] They certainly did not lose out on the Pill. In 1964 Searle made 38.9 per cent return on investment.[101] They made a sizeable 27.9 per cent return in 1968, the year that 'side'-effects were first beginning to be reported, and in which sales temporarily declined.

The popular success and scientific 'respectability' made all the difference. Doctors began to adopt birth control as a responsibility of their profession; governments came more rapidly to accept that they should finance and even organise birth control clinics; and traditional birth control manufacturers began to do scientific research and testing. London Rubber, one of the traditional contraceptive manufacturers, for a time even marketed a brand of the Pill under licence to Searle.

RECENT DEVELOPMENTS

The question of safety and unwanted effects arose quite early – but at first the glossy magazines published enthusiastic articles telling women that the Pill would keep them young, or make them more 'sexy', while the more serious papers carried headlines such as, 'No Relationship Established Between the Pill and Cancer' (or between the Pill and thrombosis; or between the Pill and permanent sterility). They were perfectly true. No relationship *could* be established, because so little research into unwanted effects had been done that it was impossible to say either way. Drill, who was Searle's director of biological research, wrote a book intended to dispel 'the traditional, often irrational arguments that the Pills cause cancer, diabetes or thromboembolic disease'.[102]

The Pill was widely tested before it was launched, but most of the tests were to demonstrate that it worked as a contraceptive, not that it was safe. In fact, the Food and Drug Administration licensed 'Enovid' in late 1959 on the strength of just 132 case histories.[103]

Later on, a majority of the tests that monitored the Pill's effects involved only women who had taken it for two years or more. This automatically excluded women who had given it up within two years. Many of the symptoms, however, show up within six months, so the safety trials excluded women who had given it up within two years as a result of the 'side'-effects.[104] A more recent study of hormonal contraception, however, has assumed that the women who failed to return for a second course of the drug had *not* given it up as a result of 'side'-effects, but must have had some other reason, for example moving from the area. The reason they gave was that women experiencing unwanted symptoms would be sure to return to the clinic![105]

Even today, systematic records are not available, either of the number of women taking the Pill (or any other drug), or of the numbers showing symptoms of various illnesses that might be related. It was eight years after the Pill was launched before the first real evidence about thromboembolism[106] was published in 1968.[107] The authors of that study and another research group, the Royal College of General Practitioners' Oral Contraceptive Study, published further reports in *The Lancet* of 8 October 1977. These studies, involving 60,000 women, indicated that the risk of dying from circulatory diseases such as heart attacks and strokes is five times greater for women taking the Pill than for women who have never taken oral contraceptives. The greatest risk is for women aged thirty-five or more, women who have taken the Pill for over five years, and women who smoke.

Until that report, which was fairly heavily publicised, ordinary women did not really know what the likely effects of taking the Pill were over a period of years. These findings, however, relate to only one group of diseases. The connection with cancer, liver damage and several other possible diseases is still not properly understood. However, the evidence has been steadily accumulating since the 1968 report. At the end of 1969 a number of brands were withdrawn when the high dose of oestrogen they contained was thought to be linked with a high cancer risk. 'Volidan', hailed as the 'British Pill' in 1963 and awarded the Queen's Award to Industry in 1967, was withdrawn at the end of 1975. In 1976 reports were published that long-term sterility might be caused by the Pill. Eighteen years after the Pill was first sold, we still await conclusive evidence about many suspected effects.

In any case, risk varies with the individual. Some women are more susceptible to headaches, nausea, thrombosis etc. – and you can't tell if you're susceptible to thrombosis until you've had it. Risk is always a relative question. All drugs have good and bad effects. It is necessary to weigh up the likely benefits and risks to the individual before deciding whether to take a particular drug. Most people, for instance, would be prepared to accept a greater risk from a cure for cancer than from an aspirin. A woman who knows that another pregnancy will kill her is more likely to accept a certain risk from the Pill than a woman who plans to conceive in a few months' time.

Since the Pill was first launched,[108] other companies have begun to market their own brands. There are now about forty-five different brands. Some were only different in name, being marketed under licence to another company; some differed in the doses of the two hormones. All companies reduced dosages with successive products, and some brands were introduced containing no oestrogen at all; while several contained new progestins, about ten different ones have been introduced:[109]

The most novel of the subsequent birth control drugs has, however, been Depo-provera, which is taken by injection every three or six months, instead of by mouth every day. Despite the undeniable advantage of this dosage form (which avoids the problem of remembering to take the Pill every day), it is outweighed by its disadvantages and, in the West, has not been licensed for general use. It has not been passed in America because there is evidence that it might cause cancer; or in Britain because there is evidence that it might cause sterility; although it has been used in clinical trials for limited periods in both countries. Elsewhere it has been shown to cause heavy irregular menstrual bleeding, or to stop periods altogether.[110]

Upjohn, the US-based multinational company that makes Depo-provera, spent a lot of money on its development. When it failed to gain approval in the United States and UK, therefore, they launched a big promotion campaign in the Third World, where drug regulation is either non-existent or not very strict. The promotion has been successful: Depo-provera has been administered to about 1 million women.[111] Doctors who take the view that it is more important to stop women in developing countries from breeding than to be cautious about their health have helped to make this 'success' possible. Dr McDaniel, of the McCormick Hospital in Chiang Mai, Thailand, for example, far from explaining the risks involved, does not (according to a report in the *Sunday Times*) even give women the standard cancer detection tests that are usual in most Western family planning clinics.[112]

Even in Britain, where the introduction of new drugs is more strictly controlled than in the Third World, Depo-provera is being used. It is licensed for short-term use, for women who have just been immunised against German measles, or where a couple is waiting for a vasectomy to take effect. There is, however, nothing to prevent a doctor from prescribing Depo-provera in other circumstances. Working-class women in particular, in the East End of London and in Glasgow, have been given injections of it by doctors who have decided they are 'unreliable' with other methods. The women themselves are rarely the ones to decide, and they are not always told about possible 'side'-effects or that the drug is still on trial.[113]

People in underdeveloped countries have frequently been used as guinea-pigs for the testing of drugs, and a large number of drugs that have failed the more stringent safety requirements of Western countries have been marketed in the Third World. Depo-provera is not the only example. The Pill itself was originally tested in Puerto Rico, not the United States. Several reports[114] appearing in the last few years have criticised drug multinationals on these grounds, and have argued that there is an enormous gulf between commercial marketing possibilities and the real health needs of people in these countries.

In the autumn of 1976, the idea of making the Pill available in Britain without a doctor's prescription – like aspirin – was considered by a government working party.[115] Many women receive their repeat prescriptions from their GPs from one year's end to the next, without so much as having their blood pressure checked, let alone having a cervical smear test done. So whether a doctor is involved in giving the Pill is not the main issue. What is important, however, is that women get their supplies, that they should have regular check-ups, and that they should know how to recognise any symptoms that may be caused by the Pill.

GPs demanded, and got, an extra fee for seeing their patients about birth control. At a conference in 1976, several GPs then threatened to remove patients from their practices if they went to a family planning clinic for birth control supplies, thus depriving them of their fees.[116]

But my dear girl, we must expect _some_ side effects!

In Britain we rely for our safety on 'advice' from a committee of experts, the Committee on Safety of Medicines (CSM). They base their recommendations about licensing new drugs on information from the company concerned, plus any research reported in medical journals. They cannot commission tests and research as required. When a drug safety committee was set up, as a result of the thalidomide disaster, it was chaired by Sir Derrick Dunlop. He then became a director of the Stirling Winthrop Drug Co., suggesting that his outlook and values were those of the drug manufacturers.

The Pill has, in fact, been more closely monitored in Europe and the United States for 'side'-effects than any other drug. The unwanted effects of ICI's Eraldin (Practolol)[117] – which caused it to be withdrawn after five years on the market – have led to suggestions[118] of there being some kind of post-marketing surveillance of new drugs. So far these are only suggestions, and suggestions opposed by the drug industry.

At last information is slowly becoming available – to doctors – about the risks associated with the Pill. But how many women read *The Lancet*? Not all doctors read it thoroughly every week; and doctors do not necessarily pass on the information to their clients. On the contrary, most doctors believe that decisions about prescriptions and risks are not matters for discussion with their patients. Properly controlled, long-term testing for all suggested 'side'-effects should be carried out immediately, and systematic records kept of the number of women taking the Pill. (The same should of course apply to all drugs.) We must know the risks involved in taking the Pill (and other drugs) so that we can decide whether or not to take it on the basis of real information.

10

Reproductive Engineering:
The Final Solution?

Jalna Hanmer and Pat Allen

Science, it would seem, is not sexless; she is a man, a father, and infected too.[1]

Virginia Woolf

This article describes current research and technological innovations in reproductive engineering.[2] We question the dreams, hopes and aspirations of those engaged in these innovations. We concentrate on the meaning these developments may have for women, not only because they are the subject of this book, but also because there are unexplored issues of importance.

Reproductive engineering has been defined as

> covering anything to do with the manipulation of the gametes [eggs or sperm] or the foetus, for whatever purpose, from conception other than by sexual union, to treatment of disease *in utero*, to the ultimate manufacture of a human being to exact specifications. . . . Thus the earliest procedure . . . is artificial insemination; next . . . artificial fertilisation . . . next artificial implantation . . . in the future total extracorporeal gestation . . . and finally, what is popularly meant by (reproductive) engineering, the production – or better, the biological manufacture – of a human being to desired specification.[3]

This definition, taken from the *Journal of the American Medical Association*, is quoted by the US Sub-Committee on Science, Research and Development of the Committee on Science and Astronautics. The aim of their two reports, one in 1972 and another in 1974, is to provide information on new research in molecular biology and techniques affecting reproduction in humans. Our definition would include the above techniques and also sex predetermination and parthenogenesis (a form of asexual reproduction to be discussed later).

While much of this research is undertaken by medically trained people, particularly gynaecologists and obstetricians, to be able eventually to 'manufacture biologically human beings to desired specifications' involves far greater understanding of how life is transmitted and maintained than we have at present. Geneticists, embryologists, cell biologists, molecular biologists and biochemists are working on altering genetic material and studying developmental processes from fertilisation onwards in many organisms from microscopic bacteria to mice and rabbits. Doctors are concerned with artificial insemination, fertilising eggs and growing foetuses outside the woman's body, developing an artificial placenta, developing products to influence which sex is conceived, and with ways of interrupting pregnancy and determining inherited and other damage to the unborn child. Thus geneticists and other scientists are working on fundamental questions of how life is maintained and reproduced, while doctors are working on processes that may immediately influence the life of individuals and society. The pharmaceutical companies translate basic discoveries and developments of either scientists or doctors into products that can be sold on a mass market. In this article we also describe the partnership between science and technology, that is the potential uses to which a greater understanding of reproductive processes can be put.

Scientists working in areas of livestock reproduction are also interested in ways of enhancing profitability. Sex predetermination will enable dairy or beef herds, for example, to be built up more quickly.[4] *In vitro* (in glass, or what is commonly called test-tube) fertilisation will allow the embryos developed from the eggs of good cattle to be carried by inferior stock, and other processes open up the possibility of producing genetic replicas of prize animals.

To ask *why* research on human reproduction is occurring is one way of exploring what such research means. There are numerous rational-sounding, even good-hearted, reasons advanced, centring around the idea of improving life for the mother, the baby and the species. One reason is that women have a 'right' to have a baby (their infertility can be overcome by *in vitro* fertilisation) or to have one of a particular sex. By controlling the genetic encounters between a particular egg and sperm through *in vitro* fertilisation, or between an egg and nucleus of a cell through cloning (another form of asexual reproduction to be discussed later), it is said that transmission of inherited disease may be eliminated. By controlling the gene pool (that is, the total number and type of genes – the hereditary material – in circulation among all people), it is said that the species may be improved. And finally, there are personal motivations of promised fame described by Leon Kass as the 'lure of immortality promised the father of the first test-tube baby'[5] and wealth for those able to market products that, for example, successfully enable parents to predetermine the sex of their children.

The scientific assumptions underlying these 'explanations', however, are controversial within the scientific community, as control on any of these levels (mother, baby, species) is problematic. One way of defusing possible criticism is to restrict the claims for the research to its immediate applications, thus side-stepping its wider implications. Edwards, for instance, justifies his *in vitro* fertilisation experiments on the grounds that some infertile women (those with Fallopian tube blockages) will be enabled to produce children.[6] He dismisses with disdain detractors who raise wider issues. Sex predetermination has been justified on the ground that it will prevent sex-linked diseases such as haemophilia. As a majority of the population approves of overcoming infertility and eliminating birth defects, we may expect people in favour of the new reproductive technologies to phrase their arguments around these issues. But this tactic obscures a potential soft underbelly of exploitation: of women and their bodies, of foetuses, and of people generally.

In an unusually sensitive article, the physician and biochemist Leon Kass describes the dehumanisation that could flow from reproductive engineering.[7] His concern with the foetus, the problem of who decides which ones are to be washed down the sink after the experiments, etc., is moving. His concern with the potential harm that could be done to the family and to humankind is also moving. But the catalogue of future ethical issues does not include the future of women. When it does, misogyny may be barely disguised. For example, Edward Grossman, in *The Obsolescent Mother*, lists the 'benefits' that will accrue from the artificial placenta.[8] These include simplifying the 'sexing' of the baby and ensuring paternity. 'For the first time it will be possible to prove beyond a shadow of a doubt that a man is the father of his children.' Motherhood, he says, will no longer be more important than fatherhood. By transferring the growth of the foetus from the woman's body to a machine, he claims that the sense of awe generated by pregnancy and childbirth will have nothing to feed on.

We must not assume that research into reproductive engineering is limited to capitalist countries; that the motive for developments in this area is simply a desire for greater profits. Grossman reports that the Institute of Experimental Biology in Russia is engaged in *in vitro* fertilisation experiments, and that the Chinese are interested in where this might lead. The Chinese Communist Party paper, *Jenmin Jin Pao*, states editorially:

> These are achievements of extreme importance, which have opened up bright perspectives for similar research. . . . Nine months of pregnancy is no light or easy burden and such diseases as poisoning due to pregnancy are detrimental to health. If children can be had without being borne, working mothers need not be affected by childbirth. This is happy news for women.

The new dimension is to gain more work from women, but the equation remains the same: women are to be made more like men by taking reproduction from them; or, alternatively, women are to be made less like women. To ask why this should be desired is to open a Pandora's box of male subterranean desires, feelings, motivations.

The reproductive techniques to be discussed in this chapter offer new future choices in social organisation. The relation between the sexes could be transformed. But in what direction? Control of female sexuality and reproductive capacities is a constant factor in human societies throughout history. Age of marriage, regulation of sexual intercourse, various forms of birth control including abortion and infanticide are commonly used forms for the control of female fertility in human societies. But in our Western Christian society control over female reproductive capacity was largely experienced in the past as forcing us to have children throughout our childbearing years, by denying us culturally approved ways of limiting births. We are now moving into an era when we will have the scientific and technological knowledge to be able to deny women the opportunity to reproduce, or to reproduce only if they use the genetic material of others. In this chapter we speculate on the direction of potential change, giving due weight for male opinions.

Male writers, including those scientists who are prepared to speculate on the future, see a restriction in the lives of women as a result of *in vitro* fertilisation techniques – literally, through a reduction in their numbers relative to men, and socially, through a diminished role in society. There are *no* male writers who make the female argument of, for example, S. Firestone, that relations between the sexes will become more equal, and thus more loving and more harmonious, once babies are made in factories.[9]

Future heightened control of women need not be overtly coercive, but could be largely sustained by ideology. For example, ideologies of female subordination and the desirability of capitalism combined could be important ingredients underwriting some variant of this market economy scheme offered by Nobel Laureate Shockley.[10] He outlines five steps to a population control plan that reinforces the monogamous unit and strengthens male control. It begins by convincing people that population limitation is desirable and necessary for survival. The Census Bureau then calculates the number of children each woman may have (2.2, if one-third of one per cent increase is permitted each year). The Public Health Department then sterilises every girl as she enters puberty by a subcutaneous injection of a contraceptive capsule, which provides a slow seepage of contraceptive hormones until it is removed. When the girl marries she is issued twenty-two deci-child certificates. Her doctor will remove the contraceptive capsule on payment of ten certificates and will replace it when the baby is born. After two babies, the couple may either sell their remaining two certificates (through the Stock Exchange), or try to buy eight more on the open market and have a third child. Those who do not have children have twenty-two certificates to sell. In this scheme control of female reproduction passes largely to the state, and, to the extent that males are superordinate in society, to men in and out of the family.

THE ART OF THE POSSIBLE

The degree of potential control that could be exercised over reproduction, and therefore over women, varies with the effectiveness of the technique. For example, sex predetermination products presently on trial will be less than 100 per cent effective, but cloning will ensure 100 per cent success. The level of technology also affects the potential for control. Reproduction will remain a cottage or craft industry until the artificial placenta is perfected, as each foetus will need an individual woman to carry it whether the egg is fertilised by artificial insemination or by cloning. Even if social forms are developed to intensify production (for example, fewer women having children throughout their reproductive years), the comparison remains on the level of battery v. free-range hens. Once the artificial placenta is developed, however, the way opens for factory techniques or 'baby farms' to become the mode of production. The elimination of women, or femicide, becomes a possibility, although until some means is found to give the ability to programme the entire development of an organism to any body cell (if it is possible to do so) some women must be kept for their eggs. Alternatively, women could be bred for particular qualities, like passivity and beauty. We begin by describing current or soon-to-be available technologies.

Artificial insemination

Artificial insemination is the introduction of sperm into the vagina by means other than sexual intercourse. Widely practised in animal husbandry – especially since the technique of storing sperm by freezing began to be developed thirty years ago – artificial insemination has been used to improve stock through the use of sperm of prize animals. Artificial insemination is rarely used among humans.[11] Individuals requesting it must conform to criteria determined largely by custom, that is, they must be a married couple where for reasons of infertility or low potency the husband is unable to fertilise his wife. The woman may be required to undergo extensive tests to prove her potential fertility before artificial insemination is tried, even though the latter is a remarkably simple act, unlike some of the other tests she must submit to. Unmarried women, lesbian and heterosexual, find it more difficult to obtain artificial fertilisation; witness the public uproar when one woman of a lesbian couple conceived in this way. Questions were asked in the House of Commons, the doctor concerned went virtually into hiding, and the press had a field day, implying that if such practices were not illegal then they should be made so. What is so interesting about artificial insemination is its slow extention to human reproduction. Powerful cultural norms and values inhibit the growth of this practice. Why should this be so?

If artificial insemination were readily available to all women who requested it we could expect an extension of its use, particularly among women who do not want to conceive in the usual way and who possibly may not wish to live with a man or with the father of the child. Such behaviour would be a potential threat to men psychologically and to their superordinate position in the family. The truth is that at present women do not need men for reproduction to the same extent that men need women: with sperm banks, few men are really necessary. Perhaps men are beginning to realise this; if so, we can expect them to be frightened and to seek to retain power over reproduction.

It is interesting to contrast the slow extension of artificial insemination from domestic animals to humans with the expected rapid growth of sex predetermination. Unlike altering the male role in human reproduction, there seem to be no cultural norms or values to slow down a potential imbalance in the sexes likely to result from marketing sex predetermination products.

Sex predetermination

Men express greater preference for male over female children than do women, although both prefer boys, and these preferences are acted upon as more families stop having children when the last child born is a boy than when it is a girl.[12] Given the reality of greater social and family power held by the male 'head' once it is technically possible, will women have the sex of their husbands' preference, their own wishes being ignored? If women always had the sex they wanted the imbalance between males and females would be lessened but greater than at present (105 male to every 100 female births).

The choice of whether to have a boy or a girl child might be achieved in several ways. The existing method involves diagnosis of the sex of an embryo in early pregnancy, while future developments are concerned with the determination of the sex of the child at conception by controlling which type of sperm fertilises the egg.

In humans the sex of an individual is controlled by special chromosomes known as sex chromosomes, of which there are two types, called X and Y. There are two sex chromosomes in every cell. In a woman these are both X chromosomes while in a man there is one X and a smaller Y chromosome. Every egg has one X chromosome, and a sperm can have either an X or a Y. If an X-carrying sperm fertilises the egg the baby will be a girl, while fertilisation by a Y-carrying sperm produces a boy.

The amniotic fluid that surrounds the foetus contains sufficient quantities of foetal cells and chemical substances to enable information to be discovered about the embryo by drawing off a sample of the fluid and carrying out various tests. This technique (amniocentesis) is used to detect cases of spina bifida and Down's syndrome ('mongolism') so that a termination can be offered if the child is abnormal. It is also possible to identify the sex of the embryo by identifying the sex chromosomes of its cells.[13] There are other methods of pre-natal sex diagnosis which vary in their reliability and the extent to which they have been tested. These include looking for foetal cells in cervical smears or blood samples from pregnant women, and they have the advantage of eliminating the small risk to the foetus involved in amniocentesis. Selective abortion of the unwanted sex is thus a present possibility.

The acidity or alkalinity technique has been written about within the Women's Liberation Movement. Vinegar and soda bicarbonate douches before intercourse timed to coincide with pre-ovulation or ovulation have been publicised as a means of predetermining the sex of the child. A refinement of this method is most likely to be available commercially soon. It is known that an alkaline environment favours the survival of Y-carrying sperm, while an acid environment favours X-carrying sperm. Dr John Pollard in Manchester, for example, basing his work on the findings of several other investigators, has developed a gel (or gels) which, when used on the vaginas of rabbits substantially increases their chances of having either male or female offspring. [14] He is looking for a pharmaceutical company to produce and promote his product commercially, once trial runs on women are completed. It is unlikely that any chemical method will give 100 per cent reliability (and some scientists think that such methods will never work at all), but 'success rates' of 70–80 per cent have been predicted, which would be adequate for commercial development. It is possible that sex predetermination products may be marketed before the publication of this book.

Some gynaecologists have been investigating suggestions that the mother's diet may have an influence on the sex of her child, for example that a high salt diet increases the chances of having a boy, but so far the samples used have been very small and the results inconclusive.

Another method is to separate sperm. An Indian zoologist noticed that the peasants in his country preferred to have their cattle artificially inseminated at dusk. They believed that this produced more male calves, which proved to be true. The container of sperm had been sitting around all day, the X-carrying sperm sank faster than the Y, so when the inseminate was drawn from the top of the container there was a high proportion of Y sperm. Later it was discovered that chilling sperm to prevent them from swimming around made separation more efficient. There are other differences between X and Y sperm; for example, Y sperm can swim faster and X sperm can survive for a longer time in the woman's body. The practical results of these properties have been known for centuries to Orthodox Jews, who consistently achieve a higher-than-average ratio of boy to girl babies by obeying the directives of the Talmud: intercourse is timed to coincide with the woman's ovulation (because it does not take place during or for one week following menstruation), and ejaculation should occur after the woman's orgasm (which increases the alkalinity of the uterus).

Scientists have been working for many years on refining techniques for separating X and Y sperm based on what we know of their physical properties.[15] No technique so far gives consistently good results, but it should not be long before improvements are made enabling efficient separations to be carried out. Considerable impetus for such research comes from the economic gains to be made in livestock breeding, but work is also being carried out on human sperm.

In addition to what we know about adult sex preferences for their offspring, sex predetermination should be of particular concern to us because there has been little public and scientific discussion of its ethical and social potentialities. For example, in the US Sub-Committee Reports referred to earlier, sex predetermination is hardly mentioned and is not seen as problematic when it is (which presumably is why it is hardly mentioned):

> Wide use of either method (aborting embryos of undesirable sex determined by analysis of amniotic fluid or by separating XX or XY [sic] sperm cells) might cause a marked imbalance in the sex ratio in the population, which could lead to changes in our present family structure (and might even be welcomed in a world suffering from overpopulation). Alternatively, new social or legal pressures might be developed to avert a threatened imbalance. But though there would obviously be novel social problems, I do not think they would strain our powers of social adaptations nearly as much as some urgent present problems.[16]

The reference is oblique; whether the imbalance is to be more or fewer females than males is not mentioned, presumably because we are all expected to understand that we are discussing a reduction in females relative to males. This is either desirable (as population will be limited) or a matter of little concern.

These attitudes are not unique to this male biologist. For example, another, J. Postgate, describes the social consequences of the reduction in females as a 'matter of fact, rather than serious concern'.[17] He suggests that women would be kept in purdah, no longer able to work or travel freely, given as rewards for the most outstanding males; polyandry could be introduced and women come to be treated as queen ants. Even more dramatic, he predicts that in the Third World population would be reduced, as 'underdeveloped, unenlightened' nations would 'leap at the opportunity to breed male'. Sex predetermination thus also becomes a form of birth control. Sociologists may be more cautious, predicting a smaller swing towards males and less social transformation, but the substance of the argument is the same.[18] Another possibility is that women become not only a numerical minority but also a nation of younger sisters. It may be that families in Western industrial countries would be more likely to have a boy first and then possibly a girl, but birth order alone can affect personality and life chances.

In vitro fertilisation

One innovation that has received wider discussion and research is *in vitro* fertilisation. The best-known work on *in vitro* fertilisation is being done in Britain by Edwards and Steptoe, and was begun in the 1960s.[19] They demonstrated that it is possible to fertilise a human egg outside the woman's body and induce it to begin cell division. This work was originally funded by the Ford Foundation, but in 1974 it was disclosed that the Medical Research Council had refused to provide further funds on the grounds that there was no evidence that it could not produce abnormal babies. (Presumably they did not want a thalidomide-type disaster on their hands.) Edwards, however, disputes this and now claims to have successfully implanted an embryo which was born in 1978. There is no reason to doubt the possibility of this, as the egg begins cell division within hours of fertilisation and the embryo does not attach itself to the walls of the uterus for several days. Preparing the woman's uterus to receive the implant involves hormone treatment and careful timing. Once this procedure is well enough understood to be easily repeated then the use of women as incubators for someone else's child becomes feasible.

Artificial incubation for longer than the first few days belongs to the realms of the future, as do cloning and parthenogenesis.

THE NOOSE TIGHTENS

Both cloning and parthenogenesis offer methods of reproduction using the genetic material of only one person. With parthenogenesis this will always be that of the mother, while with cloning the genetic material may be that of any person. Some radical feminists have expressed interest in cloning and the artificial placenta as a way of resolving antagonism between the sexes – which is seen to arise from the differential roles men and women play in reproduction.[20] Other radical feminists have expressed interest in parthenogenesis as only daughters would be produced; however, cloning, using only the mother's genetic material, would produce the same result.[21] Both these views imply a faith in technological solutions and a disregard for who holds social power and controls the state: points we will return to once we have described these techniques.

Cloning

Cloning is asexual reproduction achieved by taking single cells from an individual and inducing them to begin dividing in order eventually to produce adult organisms which will be genetically identical to their parent. But while they will have the same genetic potential they will not be exactly the same, as the environment in which they develop will have a major influence on how they eventually turn out.

Men, too, have expressed interest in cloning. It has fired the imagination of many writers, though most have erroneously seen it as a method of producing exact photocopies of living humans. It can be presented as a means of fulfilling personal dreams or ambitions. A recent book published as fact, although scientists regard it as a hoax, describes how an ageing and heirless millionaire found a doctor to produce by cloning an exact genetic replica of himself.[22] Interestingly, in this book the egg donor and the woman who carried the child to term were led to believe that they were participating in infertility research. Cloning can be visualised as the basis for a new social order, as in Aldous Huxley's *Brave New World*. Science fiction writers, and at least one eminent geneticist, have described how cloning might be used to colonise new planets, the astronaut incubating a few skin cells from his arm, for example.

However, at present among higher organisms this kind of cloning can be accomplished only with plants; it is possible, for example, to grow a carrot from a single carrot cell. In animals, it is much more difficult, but some important work was done in the early 1960s with toads.[23] The nuclei (the part containing virtually all the genetic information) of unfertilised toad eggs were killed. Other nuclei were taken from the gut lining of tadpoles and transplanted into denucleated eggs. About one per cent of the treated eggs developed into normal adult toads (the rest died at various stages of development or failed to start dividing at all). Other workers using similar techniques with frogs had different results. They found that if the donor nucleus was taken from an early embryo development could proceed, but by the time the embryo reached a certain stage of development its nuclei appeared to have lost the ability to programme the entire development of a frog.

Work is continuing on nuclear transplantation in amphibians (frogs, toads and newts).[24] The scientists involved hope that this research will help to answer important questions about how genes control development. There are also implications for cancer research and, perhaps, for the prevention of abnormalities that arise before birth. Most work has been done on amphibians, mainly because their eggs are very large and easily manipulated. Mammals' eggs, including those of woman, are very tiny in comparison, and the techniques enabling transplantation to be carried out without causing lethal damage have yet to be fully developed. However, recently a Polish investigator reported successful nuclear transplantation experiments with mouse eggs.[25] Techniques exist which enable cells to be fused together artificially, and these may provide part of the solution. So cloning mammals may still be a long way off, but there is no reason at the moment to think that it is an impossibility. Estimates of how long human cloning may take vary from several to forty years or so.

The part of the egg outside the nucleus is essential for development. So although the donor of the nucleus can be of either sex, women will be necessary to provide the eggs. Cloning, however, offers a way of eliminating the woman's *genetic* contribution to the foetus because her egg must be denucleated. If the egg cell nucleus (her genetic contribution) is replaced by that of another cell from another person, who can be either male or female, the genetic makeup of the foetus, including its sex, will be that of the new nucleus. The implications for animal husbandry, where prize animals of particular sexes are commercially desirable, are obvious; the extension of this technique to humans would involve a much greater level of control of women than currently exists.

While considerable research is being carried out on cloning, relatively little work is being done on parthenogenesis. It has also received less public and scientific discussion than any of the other techniques we have discussed.

Parthenogenesis

Parthenogenesis is the production of an embryo from an egg without fertilisation by a sperm.[26] The embryo may or may not develop into an adult. A recent survey estimates that about one thousand animal species reproduce exclusively by parthenogenesis, and still more may use both parthenogenesis and fertilisation at different times.

Familiar examples are bees and aphids, but even among vertebrates as highly evolved as lizards species have been found to consist entirely of females reproducing parthenogenetically. Among the warm-blooded animals parthenogenesis has been reported in several species of birds. Turkey populations with a high incidence of parthenogenesis have been selectively bred, but the chicks produced in this way usually die before hatching, and seldom survive to adulthood. Parthenogenetic early embryos can occur naturally, or can be induced by various methods in several species of mammals such as mice, rats, ferrets, horses, rabbits and humans. Although there are occasional reports of adult mammalian parthenogenetic offspring, including humans, none has been proved conclusively. Mammalian parthenogenetic offspring will all be female, as the mother has no Y chromosome to transmit. Parthenogenesis, therefore, offers no explanation for some of the more celebrated legendary accounts of virgin births!

In 1955 a woman scientist gave a lecture describing her research with guppies (a type of fish) and suggested that parthenogenetic births also occurred among humans. A Sunday newspaper ran the story, inviting women who thought they had parthenogenetic daughters to have their claims tested by a panel of doctors and scientists. The tests were based on the expectation that mother and daughter should be genetically very similar, since the daughter would have received all her genetic material from her mother instead of half from each parent, as is usually the case. Out of the nineteen volunteer pairs, only one passed all the tests for blood and saliva similarities and eye colour factors, but the probability of the claim was reduced when a skin graft from daughter to mother was rejected after four weeks. The study was inconclusive, as the daughter need not be completely genetically identical with her mother to have been reproduced parthenogenetically; but even so this investigation has not been repeated.

For a mammalian egg to develop into an embryo it must have the full number of chromosomes found in every body cell. In humans this number is 46. The egg is derived from a cell containing 46 chromosomes, but before it matures it goes through a division in which 23 are lost. The sperm provides the other 23 chromosomes; the fertilised egg now has a full set and development can take place. If the egg is to develop without a genetic contribution from a sperm it is clear that its chromosomes must be doubled at some stage in its formation.

The usual result of parthenogenetic conception in mammals,

including humans, is an embryo that dies before implantation. Chromosome doubling has been suggested as the reason for this. Almost everyone carries some genes that have the potential to cause gross damage and death during development, but when genetic material is contributed by the mother *and* the father these lethal genes are usually masked. Parthenogenesis involves doubling of the mother's chromosomes, and any lethal genes will be doubled as well. The daughter then gets two copies of the lethal gene, and death of the embryo results. However, if this were the only explanation we might expect inbred strains of mammals to produce viable parthenogenetic offspring, since such strains do not carry lethal genes; but investigations of inbred strains of mice suggest that their parthenogenetic daughters also die at an early stage. Perhaps the sperm makes some indispensible physiological contribution to the fertilised egg, apart from its genes. These scientific problems mean that parthenogenesis is not a viable alternative for most women who want to conceive children without men. For those few who might have the genetic potential, we do not have the technology to induce the egg to retain its full set of chromosomes and to begin cell division. This stage of technology has been reached only with animal experiments, particularly on mice.

The artificial placenta

The artificial placenta is likely to be the last to be achieved of all the technical processes discussed in this article. It is possible to maintain the embryo *in vitro* until the sixth day after conception, but not thereafter, as the requirements of the foetus are too complex and exact. These requirements are met by the placenta, which allows for the passage of nutrients into the embryo/foetus and passage out of waste products, as well as providing an intricate feedback system of hormone balance. Medical scientists have been working on the problem of maintaining premature foetuses at earlier and earlier stages in their development; and other scientists, in Britain and the United States and possibly elsewhere, are attempting to develop an artificial placenta capable of carrying the child to full term from fertilisation.[27]

Edwards reports that:

> the first six days of embryonic development of the blastocyst stage can occur *in vitro* . . . mid-term abortuses have been maintained in culture for a few hours or days . . . and premature human babies can be in-

cubated from 24 weeks of gestation. . . . Almost one-half of pregnancy is thus replaceable *ex vivo* [outside the womb].[28]

The technical problems are so great, however, that replacing the other half of pregnancy outside the womb is unlikely to be accomplished quickly. Thus women are likely to be needed both to provide eggs for fertilisation and to carry foetuses for some time.

SEIZING THE MEANS OF REPRODUCTION

A question that cannot be ignored: is reproductive engineering just another attempt by males to control the sexuality and reproductive capacity of women, but this time by wresting control completely from us? Forced reproduction, and reproduction denied, are two sides of the same coin.

Radical feminists have seen parthenogenesis as the area in which work should be done as it would give women control over reproduction. 'Seizing the means of reproduction' is seen by many women as basic to the feminist revolution. This can be feared by men, as women who wish to avoid them could reproduce girls and 'pass on to their daughters an ever-mounting disdain for men that in several generations could lead to the establishment of matriarchy, with eventual demand for separate statehood'.[29] How general this doctor's fear of female independence and control over reproduction is remains unknown.

What is known is that female reproduction is at present almost entirely in the hands of males, from the males who make the birth control devices to the male doctor who delivers the baby or does the abortion. Reproduction engineering, too, is almost entirely in the hands of men, as are the governmental and voluntary agencies that fund this research and the decision-makers in the companies that will commercially exploit these scientific findings, reaping large profits. Further, the question of whether women want these methods of artificial reproduction will never be put to us, and certainly the decision about whether to go ahead or stop will not be ours so long as society as we know it exists.

The reality is that men live in harmony with women by subordinating them. Control is generally exercised over women's sexual and reproductive capacities through customs that determine when and how we may have children, and through the appropriation of these children through kinship, inheritance and marriage customs. In our society our sexuality is often devalued and denied. An ideology justifying the general devaluation of women is essential if material exploitation is to succeed; it lays the basis for adherence to social customs and rules that enable men successfully to exercise individual and collective power and authority over women. Ideology is always backed up by force and the threat of force, and in our society women are controlled in part by a fear of random violence in the home and public places.[30] To be physically and sexually abused, and even murdered, is the fate of many and the fear of many more.

The relation between socially sanctioned acts of violence against women and the sexual and reproductive elements of the supporting ideology may take many forms which vary in their severity. At their most severe forms of control involve sexual assault or mutilation or even death for a proportion of the female population. In the past a major mass physical attack on women by men was directed at women as reproducers and sexual beings. The witch hunts of the thirteenth to eighteenth centuries in Europe were largely aimed at women, who are estimated to have been 85 per cent of the nine million victims mutilated, tortured or burnt alive and in other ways murdered.[31] One sure sign of a witch was the ability to procure an abortion. The witch hunts coincided with the introduction of a new social order in Europe, one based on a transformation of science and technology.

The violent control of women is exercised both within and between generations. Males may directly control females of the same generation; between the generations their control may be indirect, with elder women enforcing compliance on the younger. Infanticide is an example of a practice that might more correctly be called femicide. Societies systematically practising infanticide almost always did so through the exposure of female infants.[32] One influential explanation within anthropology of the practice of infanticide is that food shortage necessitated it; a more interesting and, we would argue, valid way of exploring the individual and social meaning of infanticide is to look at the position of women in the resulting social order. For example, polyandrous societies, where a number of men are married to or have sexual relations with one woman, are not directly comparable with polygamous societies, where men compete with each other for women. Women in polyandrous societies are not in competition with other women for men, some women doing without while others have more than one. Polyandry, when linked with female infanticide, is a severe form of control of women by men, as the biologist Postgate, in his vision of the future discussed earlier, recognises.

In our society today, to the extent that 'failure to thrive' (the term used when infant birth weight does not increase at a normal rate) is the result of parental neglect, it can be seen as a less spectacular form of infanticide. And, to the extent that failure to thrive is sex-linked, it is a less spectacular form of femicide. To explore this we need an analysis of infant mortality rates by sex for countries possessing such statistics. Third World countries provide examples of under-nourished female and well-nourished male children; where what food there is goes to the boys of the family.

We know that men attack female genitalia, controlling women through sexual assault in both state and non-state societies. [33] To-day an estimated 25 million women in Africa, in both rural areas and cities, continue to be sexually mutilated through so-called 'female circumcision'. This consists of several practices, primarily clitoridectomy and infibulation. With clitoridectomy the clitoris and labia minora are removed. In the West, while clitoridectomy was *widely* practised only among one small Russian sect, it was used as a medical cure for sexual excitment in women, with the last reported case in the United States in the 1930s. In Africa today clitoridectomy is performed on girls and adult women by women. Infibulation often is performed on young girls by their mothers and includes clitoridectomy and further excision of part of the labia majora. The two sides of the vulva are then closed over the vagina except for a small opening to the rear to allow passage of urine and menstrual blood. Infibulated females have to be cut open to allow intercourse and further cut to deliver babies; they are then resewn. [34] The many medical complications and the severe suffering of women and young girls (four to ten years of age) is rarely acknowledged by the male medical profession of their countries. Almost no research occurs.

The usual explanation for clitoridectomy and infibulation is that it controls the sexuality of women, thereby ensuring virginity and faithfulness. Women cannot find husbands unless they have been genitally mutilated. Another 'reason' given for infibulation is that it ensures fertility, but the exact opposite is true: infibulation and clitoridectomy ensure birth difficulties for mothers and babies and subsequent medical problems for those women not rendered in-fertile. [35] We should not assume that 'the natives' are too stupid, or whatever, to recognise this; that they do not admit their knowledge is another matter.

Another major permanent physical scarring of women in world history was footbinding in China. This custom lasted for a thousand years, affecting billions of women. This too had sexual overtones, as footbinding was believed to increase the folds in a woman's vagina, thus increasing male pleasure.[36] The mother bound her daughter's feet, progressively tightening the bandages that held the toes under the soles until the feet broke in their growth. The perfect foot was three inches in length and a source of sexual excitement in itself. And, of course, it is easier to keep tabs on a crippled woman, thus controlling her sexual and reproductive capacities.

While we do not have information on women successfully repelling the cultural codification of physical attacks on them by men, we do have knowledge of the part women have played in the perpetuation of these social customs. It was mothers and grandmothers, for example, who bound the feet, and who continue to cut off the external genitalia, of their daughters and granddaughters. Women have acted, and do act, as agents of male individual and social power. Social and personal necessity, while important in explaining individual compliance, does not seem a sufficient explanation for the rise of these customs; nor does ideology. While it is painful to raise, we need to further our understanding of how physical attacks by women of one generation on that of the next can be resisted. Once men and women have the option of sex predetermination we will have an opportunity to explore this issue, as women will participate, even if not completely freely, in individual decisions to have boys or girls.

Some women, particularly radical feminists, seem to accept that men intend to eliminate us. For example, Robin Morgan:

> The war outside, between women and male power, is getting murderous; they are trying to kill us, literally, spiritually, infiltratively. . . . If the political is solely personal, then those of us at the barricades will be in big trouble. And if a woman isn't there when the crunch comes – and it is coming – then I for one won't give a damn whether she is at home, in bed with a woman, a man or her own wise fingers. If she's in bed at all at that moment, others of us are in our coffins. I'd appreciate the polarization now instead of then. . . .[37]

Another example comes from the increasing entry of transsexuals into the American Women's Liberation Movement. The assumption of one article is that transsexuals want to participate in defining and creating women's culture preparatory to dispensing with women entirely. Lest this seem unduly paranoid, this quote from a US transsexual puts the issue in perspective:

> Genetic women are becoming quite obsolete, which is obvious, and the future belongs to transsexual women. We know this and perhaps some of you suspect it. All you have left is your 'ability' to bear children, and in a

world which will groan to feed six billion by the year 2000, that's a negative asset.[38]

Male envy, hatred and fear of women is dangerous for us because men have the social power, and soon the technology, to implement the 'final solution'. Of course there will be rationalisations: it will be kinder to women to relieve them of childbearing; it will be better for society to have only certain kinds of people. Sex, various abilities and physical attributes will all be graded for desirability. In this scenario, as in our society today, women are not really believed to be necessary except for child care, housework and the sexual gratification of men. Society could do with far fewer. Or alternatively, if women could be made more passive and dependent, a true slave class could be created in which men could have as many as they wanted. Once adults can be turned off the assembly lines then sexual gratification remains the only use women will have for men, as housework could be split off for a sub-human species or mechanical robots. And women should not assume they will be kept around for sexual servicing, as transsexuals could fill that role.

As mankind would continue, should women mind if they no longer exist? Would we not have played the 'helpmate' role to its zenith? Should not women accept femicide as the highest expression of femininity? If women readers feel an internal objection rising in them at the thought of this outcome, does this mean women are a separate group from men and one united in rebellious thoughts?

RESISTANCE

How can we translate these rebellious thoughts into rebellion? We have no answers, but would like to suggest that women need to involve themselves by finding out what is going on. Feminist women in science are needed to inform women continually of scientific and technological developments. We need more women from the Movement to go into science and technology. It is in the pharmaceutical companies that scientific findings are turned into technologies, where trial-runs for effectiveness are carried out, where advertising firms are contracted to 'sell' the resulting product. We must take scientific research and technological developments out of the hands of men.

We need to refine our analysis of the state as the repository of male power. We need to develop our understanding of the relationship between science, technology, the state and male power. Women need networks of Movement women discussing, monitoring, working out means of affecting decisions, possibly infiltrating key laboratories and government departments such as the Ministry of Defence as laboratory technicians, secretaries and cleaners, if not as scientists and top government officials. It is in the various departments of the state that the initial funding of projects often takes place and the state seal of approval is given to developing technologies.

Reproductive engineering offers a vehicle for the total control of female reproduction. Reproductive engineering offers a vehicle for the final working out of the antagonism between women and men. In and of themselves these potential developments need not be oppressive to women; it is the social meaning they may be given and the way in which they may be enforced on the population that is potentially oppressive. Before coercion will come the con trick, the argument, of how it is really all being done for women's benefit. Political work around reproduction must begin to include these future possibilities. It is no longer adequate for the Women's Liberation Movement to focus primarily on abortion and secondarily on issues around childbirth, such as home versus hospital deliveries: there is a knockout punch coming from another direction.

11

The Masculine
Face of Science

Ruth Wallsgrove

We live in a society divided into two unequal halves: women and men. We are taught – and indeed in our society it is true – that women and men are different. There are two sets of characteristics, femininity and masculinity, and a 'real' man is masculine, and a 'real' woman is feminine. Masculine characteristics are, loosely (because they vary from place to place): aggression; courage; strength; independence; the ability to be logical; dominance. Feminine characteristics are: loving and caring; selflessness; concern for people; intuition; irrationality; passivity; weakness; dependence; inability to put things in their (detached) perspective. . . . Now, I am *not* going to shed tears over men's inability to cry; nor argue – because to me it is patently obvious – that no real person is totally masculine or totally feminine.[1] It's clear that these characteristics are more social ideals than biological norms.

As a woman with pride, however, I have always bristled at these concepts of masculinity and femininity, because masculine characteristics are the ones most valued in *people* in our society – most of the feminine ones are greatly undervalued.[2] Even those qualities that we feel to be morally good – such as loving and caring – are ridiculed, understood by all of us to be good things when possessed by someone else, but felt to be weaknesses in ourselves. I can argue, of course, that I can be as independent and objective as any man . . . or in the same bristle point out that the world needs rather less aggression and rather more caring anyway. Both are perfectly true. As far as we know, most men aren't innately any more masculine than women; and there is nothing so great about masculinity anyway. We could go on endlessly trying to prove these two points, but there remains a question. How is it that men managed to lay claim to the valued qualities?

Apart from being valued by our society, there is something else that most masculine characteristics have in common. They are to do with power, with control – control of oneself and others, and avoidance of control by others. Not just physical control: logic is controlled thought; 'objectivity' is the ability to control one's emotions. They are all qualities of power. And feminine qualities, as taught to little girls, are qualities of powerlessness and confusion.

The philosophy of our age is scientific rationalism, and it is clearly masculine. It is objective, logical, independent, brave. Scientific rationalists argue against the suppression of truth by brute force – even as they build bigger and better weapons. But while they don't advocate husbands battering their wives – while scientists and rationalists are not valued, or indeed renowned, for their crude machismo – nevertheless rationality and its ultimate refinement, science, are undeniably masculine. Science is not a womanly subject, as girls are told all the time; and if we don't take the hint early that it's unbecoming for girls to enjoy science, we are still unlikely to get research posts.

We can argue about science as we have argued about masculinity: on the one hand, women can be, and sometimes are, great scientists[3] (there are no biological reasons why we shouldn't be); and on the other hand, science is not all a blessing (the technologies it has given birth to have harmed as well as helped humankind and the world). Women can be scientists, but do we *want* to be? This is a sticky question for feminists but, as the rest of the book shows, it may be the most urgent question facing us. Science isn't just an interesting field that we want to be allowed into; it isn't just good jobs. Science is also to do with power.

Since the time of Francis Bacon in the seventeenth century, the prevailing view within Western science was, if you know how something works, you can control it.[4] In fact, this is why he advocated trying to understand nature – not as the pursuit of truth for its own sake, and certainly not as appreciation, but as the means by which to dominate nature, to change it to our advantage. Science is not an esoteric philosophy: it's very practical. Whoever controls the laboratories – that is, whoever decides what is studied in them and how discoveries are developed technologically – has a lot of power, not just over nature but sometimes also over other people.[5]

I don't want to argue here about which particular groups of men in our society – multinational companies, governments or (less likely) scientists themselves – control the laboratories. This is written about at length elsewhere.[6] It is clear, however, that science is a powerful tool, which we don't want to leave in anybody else's hands. It's also clear that at the moment it is in the hands of men as opposed to women, and that men do not always use it in the best interests of women. Therefore the partial answer to the question, 'should there be more women scientists?' must be that we need as many feminists in science as possible at the moment.

But is science *merely* a tool – even a weapon – that men have appropriated, a tool we wish to reclaim and use ourselves? Was science just lying around waiting for men to pick up and use to their advantage?

I don't find this easy to think about, because I was brought up in the scientific, rationalist tradition that teaches that science is the triumph of truth over ignorance and superstition. Philosophies are not things you throw off easily. When I think of the possibility that science isn't plainly and simply *true*, I get nervous, no matter how much I learn about other ways of looking at the world, even though I can see from history that *whatever* scientists believe now may well be proved wrong or irrelevant, or true only to a limited extent by others in the future. I know that scientists are not value-free – they are imbued with the values and ideas of the society around them, like everyone else, and carry them into the lab with them. Of course the scientific body of knowledge can only be approximately and limitedly 'true', and will tend to reflect how society in general looks at the world. But a nagging question remains: isn't the scientific way of thinking still the clearest way of thinking?

I want to approach this question – sneak up on it – by going back to a different one: why is science defined as inherently masculine, given that we've little reason to think that men are biologically more scientific, logical or objective than are women? The *result* of its being defined as masculine is that it both frightens girls off from science, and justifies the fact that men control the laboratories – a neat trick, but not one that was thought up by some ambitious male scientist: it goes back much further than science.

What do 99 per cent of politicians, monarchs and dictators, company directors, company chairmen, most lawyers and doctors, owners of property and controllers of finance, most heads of schools and local councillors, famous painters, film directors, philosophers, influential political thinkers and advertisers – in other words, those who control government, business, money, culture, even thinking –

have in common? They are *men*. No matter where you look in our society or around the world, the people in positions of institutionalised power are, apart from remarkably few token or extraordinary women, men.

Power is not simply maintained by physical force, though of course it often comes to that. Power can also be maintained by definitions, by language, by *ideology* – a view of the world that explains and justifies the inequality and stops people from clearly seeing and describing (let alone changing) what is happening.

To justify oppression it is necessary to argue that the oppressing group is superior to the oppressed group; that they are different and better. 'We deserve to be tops because we're nicer/stronger/ whiter/better at growing potatoes', to put it crudely! But this is not enough. To keep hold of power it is also necessary to discourage the desire for power in the oppressed group, and to encourage the desire for power in the group that oppresses. Qualities of power must be ascribed to the rulers, and of weakness to the ruled.

The concepts of masculinity and femininity do just that. Masculine qualities are seen as superior to feminine qualities, just as men are seen as superior to women. In every society dominated by men, whatever men do is valued more highly than what women do – *whatever* it happens to be that men do in that society. But 'masculinity' is also the system taught to boys to prepare them to take power; 'femininity' is taught to girls to stop them grasping power. If you're truly 'feminine' you can't work out what's in your own best interest, let alone fight for it. And of course, you don't want, or even know, what's in your own interest, because you always think of others first.

Science is power, so science is defined as masculine.

There is, however, another edge to masculinity, and ideologies of oppression in general. Liberal feminism has it that women are incomplete because we are deprived of education and opportunities, and because we are stunted by our training as girls; confused and put down. But the curious thing is that when we as women come together to talk about our experiences it often seems to us that it is *men*, not women, who are incomplete; men who seem to miss whole areas of life, especially in their relationships with other people. Of course, some radical women have concluded that men are *fundamentally* lacking, but I think it is unlikely to be innate! I think there are reasons why the controlling of another group of people leads to blind spots in the controllers.

To maintain power over other people it is necessary to argue that you are different and better. This can be based on the real differences between you, or on qualities you make up or pretend to have. But whatever the differences, the result is going to be that you end up despising, even fearing, the oppressed group. You despise their qualities, real or attributed, to justify treating them as inferiors. But of course ideology is not a rational affair, not necessarily even consciously created in cold blood. It is not invented by each individual who benefits from it – it is taught to them, too, by their society. If you are an oppressor you don't *pretend* to despise the qualities of the oppressed; you really end up accepting that they are despicable. And you fear any signs of those qualities in yourself – they are signs of weakness. And perhaps worst of all, you end up fearing the people you define and control because you lose any understanding of how they really are.

I want roughly to sketch out, in the shortest shorthand, three characteristics of science that are associated with this.

A VITAL QUALITY COMES TO BE DESPISED

Wanting power over other people is not compatible with caring about their real needs, or with feeling for them as fellow, equal, human beings. It's clear to me that these are totally opposing impulses. All oppressing groups, including men as a class, are likely to become selfish, contemptuous of people in the oppressed groups.

But there is something more to the lack of concern for others as a masculine characteristic. It is 'feminine' to be loving and caring, and this undoubtedly impedes women's struggle for control of our own lives – we know we should always be thinking about the welfare of other people, always worrying about how much it hurts those around us when we ask for even the smallest right, always thinking we should be fighting someone else's battle rather than our own. But many men seem alienated from caring for others, sometimes to their great distress. I believe that in controlling women for material benefits men have come to despise the one indisputable difference between men and women – the reproduction of human beings in women's bodies – as part of the ideology of male supremacy. Reproducing children necessarily entails caring and a respect for life; and in despising reproduction men have come to be contemptuous of caring, and careless with life.[7] The ultimate in contempt for life is war: it's very 'manly' to kill without anger, without even noticing.

Women not only are held to care for their children, husbands, and humanity in general, but also are believed to be somehow more in touch with nature, because of having babies, menstruating in phase with the moon, loving animals because they see them as babies, having motherly intuition about danger. By contrast, the ideology of masculinity tends to suggest that masculine men are detached and objective, rather suspicious of nature, finding it a challenge.

Modern science is a development of this detachment: it stands back from nature, seeing it only as a challenge. It demands that you overcome your emotions, or your appreciation of animals, or the rest of nature, or even other people, so that you can observe them coolly, if necessary take them apart to see how they're made. Scientists tend in some of their work to see animals and humans as machines. This may not turn out to be a reasonable analogy; but they have in the past treated them (and some still do) as though they were machines which feel no pain, dismissing our outrage at the suffering they have caused animals as illogical 'anthropomorphism' – that is, seeing us as identifying with the animal instead of standing back and noting down its reactions.[8] While science is not necessarily cruel, in the name of objective truth, it is often oblivious to the uses to which it is put. Discovering 'the truth for its own sake' is sometimes more important than the damage done by the misuse of knowledge or the use of half-knowledge along the way; damage to people and nature.[9] Scientists can be made politically aware of the consequences of their work, I suppose. The priorities of scientific research could perhaps be radically altered from preoccupations with war and profit to those that benefit the world.[10] But science is essentially a philosophy of detachment from the rest of the world of which we need to be wary. It needs to be seen as a tool towards *creating* the society and the world we want – *not* as something that *defines* the kind of world we should want or will be given.

The dangers of thinking you can stand back from the world have been made clear to us in the last decade by the ecology movement. Many people now have some awareness of the idea that we should work *with* nature, rather than trying to rid ourselves of any part of it that gets in our way. We now know more about how, by upsetting the finely balanced eco-systems of the world, we not only wipe out animals and plants, but also poison ourselves and make barren the land and the sea.[11]

I don't know how much women are, or could be, more in touch with nature; we are all deprived of many of our abilities in this society. But it's time to force our culture to re-examine and revalue women's traditional 'sentimentality', women's 'feminine' belief that the quality of life has priority over economics or efficiency or 'rational' planning.

CISSINESS

If you are an oppressor you come to despise the real or attributed qualities of the people you oppress, and put your qualities – real or appropriated – above them; they become the qualities that are most valued in your culture. What if you recognised that you had those very qualities that you were despising in the oppressed group? What if anyone else in your group noticed that you were acting in one of those despicable ways? Grown men, as they say, have been known to turn pale at the thought.

Masculinity is a clear-cut example of this. The worst insult you can throw at a man is to accuse him of being womanly; in fact, as many women ruefully note, there isn't any way of accusing a man of being too manly – for example, to get at a man who hassles us (in a most masculine manner) on the street we would have to accuse him of being impotent or a pooftah, of *not* being masculine. Men fear the appearance of anything feminine in themselves, often obsessively. (The same is not true of women: in some circumstances it is an insult to be accused of being masculine – lesbian, hairy-legged – but often it is a *compliment*: 'she does that job as well as a man'; 'she can think like a man'.)

There is no evidence that men are actually more logical, or less emotional, than women, *even in a society* that rewards their logic and punishes their displays of emotion from birth. It is strong for men to be logical and weak to show emotion, society tells them. So what happens to men's emotions, their subjectivity, their intuitive thinking, when men fear their very emergence in themselves? They become repressed; when these qualities do appear, they are called something else, or they are denied absolutely, often in a most hysterical and illogical way. Men often seem genuinely unaware of what they do; as 'masculine' men they cannot be behaving in 'feminine' ways, and so they are sure they are not.

What scientists have believed of themselves up until now has been an extreme version of this dishonesty or unawareness. Crudely, the 'scientific method', the way great scientists work, has been presented to science students as a step-by-step logical process, a process of observing the world through neutral eyes, making logical hypotheses, testing them out exhaustively – and abandoning hypotheses without a single regret at the first sign of results that contradict them. Several philosophers of science – notably, Thomas Kuhn – have considered that this is only half the story. No one, not even a newborn baby, has eyes that see the world in a neutral and unordered way; the fact is that hypotheses are not immediately rejected if they otherwise fit in with the general body of hypotheses (that is, the general body of knowledge that is called science at any moment of history). Scientists, like everyone else, work within a framework: it's not that they don't see certain things, but that they give a certain importance and interpretation to what they see according to theories already in their minds. Things can be observed, but have no meaning until a theory explains them. If something is observed that contradicts the existing framework, scientists are unlikely instantly to throw over *all* they've ever been taught or have found out, all that everybody else around them believes. In this situation, scientists are as likely to find fault in the observation as in the theory. (There may, of course, be another, simpler, explanation than that of the theory being wrong.)

It has always seemed to me no great problem to admit that science is not all logical; that it proceeds by intuition and analogy, in fits and starts; that there is a generally accepted body of scientific knowledge which would require a great deal of good evidence to contradict; and that scientists are emotionally involved with their work. [12]

Scientists have to work on hunches, and gamble on the best line to follow on the basis of things 'feeling right', or fitting in with other things, or even looking nice! If they don't bet on their hunches, they won't get anywhere; if they never admit it's anything but logic guiding them, they will find themselves having to defend everything they've ever said. And so it goes on.

It is said by people who live with scientists that there is nothing more hysterical than a scientist contradicted; and this at last makes sense to me. You can't be rational if you pretend that everything you do is rational; if you don't examine and come to terms with what you feel, your feelings will interfere anyway, but in a hidden and uncontrollable way. Science, and masculinity, are one-half of an exaggerated polarity, a dangerous imbalance. [13] There *is* no clear line between 'feelings' and 'thoughts', and you don't get one without the other.

The most devoted logician will admit that there is no such thing as logical argument on its own; argument always proceeds from premises, from assumptions made before you start. If you start from false premises, a logical argument doesn't lead to truth; and a conclusion is not good if the premises are evil. Physics and chemistry are relatively straightforward sciences in the sense that the assumptions made are not about people or lives; but biology, psychology and sociology – in decreasing order of respectability as scientific areas – all make assumptions about living beings, even about cultural values, and these must always be open to question. They are not necessarily 'logical' assumptions – they start from whatever scientists have been taught about life and culture in their particular society.

We are constantly sold ideas on the basis that they are 'scientific', and therefore nothing to do with politics (like the idea that racial differences in IQ are largely inherited);[14] but even if the evidence is collected scientifically, there is no truth or relevance to an argument unless the truth or relevance of its starting point can be argued. If an idea starts off with value-laden assumptions that are appalling in human terms, that idea is appalling no matter how thorough the statistics. Take the IQ debate. IQ is a dubious measure of anything but the ability to answer the specific sorts of questions on IQ tests; and those questions themselves are, to a large extent, based on white, male, middle-class experience. The evidence on the IQs of women and blacks is conflicting.[15] But even if all this were not true, I would still profoundly disagree with anyone saying that IQ is a measure of people's *right* to have professional or managerial jobs – in other words, to have power over other people – because that is a moral and political judgement with which I disagree. It can't be argued scientifically, because it can have no 'objective' truth.

But scientists, caught up in the scientific vision of a world reducible to logic, seeing themselves as St Georges in the struggle against irrationality and emotion, find it painful and hard to accept this – I know, because I've been there too.

FEAR AND LOATHING

In controlling someone else, you cut yourself off from that person; in imposing your definition of that person on her/him as a limitation, you lose your ability to see what that person actually is. The price all oppressors pay for their power is losing any understanding of the oppressed. Women and blacks are thought to be more mysterious to white men than white men are to women and blacks.

Of course, while white men do all the saying, the tendency will be that any views differing from a white man's will be seen as 'abnormal'.[16] But the differences between dominant men and those they dominate don't seem strange to the oppressed, because white men's values and perceptions are the explicit ones – theirs is the culture on the surface of our society. Other perceptions, other values, are hidden, never put into words, because the words themselves belong to men.

In our society, a woman, especially if she has any ambition or education, receives two kinds of messages: the kind that tells her what it is to be a successful person; and the kind that tells her what it is to be a 'real' woman. We are caught in the double-bind, the ambivalence of wanting to achieve something as a person while remaining womanly, because to want to do certain things (like flying a plane, getting a proper education or having an independent income) is seen to be wanting to be a *man*. We are in conflict; but we understand both masculinity and femininity. Masculinity, as I have said, is a positive model held up in front of women as well as men. But men are, on the whole, less ambivalent, because they do not necessarily experience any conflict. They can be men *and* people, because these are defined as the same thing.[17] Femininity is despised in boys, and so men grow up without experience or understanding of femininity. And by not seeing women as real people, they often don't see women at all.

Masculinity is straightforward, clearly defined. But the way men see femininity – that is, their idea of what women are – is chaotic.

'To be a woman is something so strange, so confused, so contradictory, so complicated . . . that only a woman could put up with it. . . .'[18]

As Simone de Beauvoir put it, men's idea of woman is that she is:

an idol, a servant, the source of life, a power of darkness; she is the elemental silence of truth, she is artifice, gossip, and falsehood; she is healing presence and sorceress; she is man's prey, his downfall, she is everything that he is not and that he longs for, his negation and his *raison d'être*.[19]

Women are defined as weak and passive – but also, at one and the same time, as terrible and powerful. We are witches and devils, as well as wives and mothers. Men tend to believe that women aren't scientific, but also that women are somehow 'supernatural'.

Women – on call twenty-four hours a day for others, rarely acting on our own, soothing and patient and without any formal power – are still objects of fear. We have to fear violence and aggression in men, because we can get beaten, raped, abused by them; but in our inarticulateness and deference we are feared. We *are* the unknown: they don't know what we think, what we really want, what our potential power is. Who knows what we might do if we ran our own lives?

I think men are caught in a vicious circle. It started with the first division to their benefit – the division of work between women and men in societies before the invention of agriculture. However that division came about, and whatever the exact form it took, it seems likely to me that by the time humanity discovered it was profitable to grow crops and domesticate animals for food, men were in a position to take control of those crops and animals, if only because women were, in general, less mobile because they bore children. From there, men have come to control the world in many ways, and even to understand it to a degree. But at each step they are also increasingly afraid of what they can't quite control or explain.

Men have sought to impose their own order on the world, to break it down into pieces that they can understand and manipulate. But all the time, out of the corner of their eye, chaos looms. Most of what they try to predict remains stubbornly unpredictable, and all the time they are chasing after smaller and smaller bits of truth.[20]

Logic is a defined way of thinking; step follows step, according to rules. It's an attempt to make order of the apparent chaos of our thinking. Science consists in breaking things down (analysis) and labelling them (classification), combined with logic; trying to tidy the world away into little boxes. These are undoubtedly very powerful methods for dealing with the world in certain circumstances. But the problem is that, the more you try to put everything into neat categories, the more uncontrollable things seem when they won't fit. The more you try to force them to fit, the less you understand their total nature – and the more frightening it seems when they act unpredictably. And so the more desperately you have to try and force them to fit. . . .

To put your entire faith in science, and to believe that everything will one day be explained logically, makes it a great deal less easy to cope with certain things here and now – like life and death, change, unpredictability. It makes a lot of things, which could be grasped in other ways, seem stranger and wilder than they need.[21]

The belief that science can answer every question is an unproven – and unprovable – assumption. Nobody has any idea how one reduces politics, morality or aesthetics to scientific questions.[22] But even within 'respectable' scientific fields there is a problem. Basically, science is *analysis*; that is, it isolates individual factors in situations that inevitably involve the interactions of several, or many, factors.[23] You investigate these individual factors by trying to hold all but one variable constant. For example, if you want to determine scientifically what a certain drug does, you have to get two groups of people who are similar in as many ways as possible – age, size, health, past medical history, eating habits (you can *never* know for sure in advance what differences might prove to be crucial[24]) – and give one group of people the drug. You have to give the other group something they *think* is the drug (so that both have the same attitude to the experiment and what they're taking); and you even have to make sure the person handing out the pills doesn't know which is which, in case they give it away to the experimental groups by the tone of their voice – we *know* this can be a vital difference.[25] Then, if the first group comes out in blotches and the second doesn't you can be pretty sure the drug had something to do with it.

But it all becomes much more complicated when something has varying effects depending on its combination with something else. It usually takes much longer to work out what's happening. For example, if the drug caused blotches only in people with red hair, you'd only discover it first time round if you (1) had some redheads in your experimental groups, *and* (2) guessed that this was the relevant factor. There are a lot of factors you might not think of checking, or even know about. If it was interaction with another drug for which you hadn't tested, you might realise only years later when somebody who had taken both had an unusually high number of cases of blotches, and started to get suspicious, and had the time and wit to pursue the matter. A doctor who knows she or he's prescribed both drugs to all the patients with blotches could make the connection. But what if it were the interaction of the first drug and, say, cucumbers, and nobody has a record of what people eat? What if the blotches were caused by having taken the drug five years previously, and then eating cucumbers?[26]

We can learn, slowly, how one factor interacts with another; we can test variations of the first against several different levels of the second, for example. There are ways of testing the interactions of three variables – it's not much fun, but it's possible! But it's not simply twice as difficult to deal with the interactions of two factors as with the effects of one: it's many times more difficult. The difficulties of dealing with a lot of factors increase astronomically with each additional factor. There is no way we could ever guess what was happening if blotches were due to having taken a drug five years ago at the same time as eating cucumbers and, say, dyeing your hair red and going to Spain. Scientists have to work in the hope that we can reduce everything to a small number of factors that matter.

Science, in other words, can isolate factors, but it can't necessarily put them back together. That's why we ought not to experiment with the more dramatic applications of science – nuclear power for instance – because *nobody knows* how they will affect us ecologically or in any other way, in the long run.[27] Such aspects of science could affect us in a thousand complex ways we can't even begin to imagine. Remember thalidomide? How many other drugs are inadequately tested before marketing? And what about long-term effects, such as the effects of diethylstilboestrol (DES), which was given to women in pregnancy, to prevent miscarriage, and was later found to cause vaginal cancer in the *daughters* of mothers who used it?[28] Sometimes the unknown risks may seem justified by the known benefits. But often – especially when the financial benefits go to the few who own companies, and the physical risks are taken by the majority who don't – the gamble isn't worth it. Scientists don't help by concealing their ignorance, or believing that science can do more than it can.

There are many situations in which it is extremely, if not impossibly, difficult to hold steady all the variables but the one you want to study. You can't stop all processes except one in a living human being, or even in a living amoeba. And we know almost nothing, scientifically, about the mind, because it doesn't appear to act simply as a number of independent processes; it's extremely hard, even conceptually, to isolate bits of thinking. Unlike, say, the effects of arsenic on the body, it's impossible to say that any one particular outside factor affects the mind in a particular way.

Science can go on finding out more about individual factors, more about their simple interactions with each other. But whether knowing a great deal more about individual components will ever lead us to understand a whole system remains an open question.

And if it doesn't, in the meantime, we've got to have some other ways of looking at the world. And some way of breaking the vicious circle of fear and dominance; of stopping those who have the power from killing us all in their panic to put us into little boxes.

Appendix:
Some Biological Information

SEX HORMONES AND THE MENSTRUAL CYCLE

Hormones are substances that are carried in the blood to various parts of the body, and have specific effects on certain parts of the body. Most of them come from specific glands (the endocrine glands), such as the ovaries, the testes, the thyroid and so on. Because they are secreted into the blood they are carried to all parts of the body, although we can usually specify a 'target organ' which is particularly sensitive to that hormone. For example, the uterus (womb) is especially sensitive to the hormones from the ovary, although ovarian hormones do affect all parts of the body, including the brain, to *some* extent.

The so-called 'sex' hormones fall into three groups, which are:
1 androgens, which predominate in the male;
2 oestrogens, which predominate in the female; and
3 progestins, which also predominate in the female, and are the 'pregnancy hormones'.

All three belong to the group known as *steroid* hormones (the name refers to substances having a particular kind of chemical structure).

All three types of steroid hormone are manufactured by the ovary, the testis *and* the adrenal glands (which are situated on top of the kidneys). What differentiates female from male is the relative amounts of each type produced: ovaries secrete mainly oestrogens and progestins, with a little androgen, whereas a man's testes produce mainly androgens, plus a little oestrogen and progestin. In other words, no *one* hormone is exclusive to one sex or the other. The adrenal glands of both sexes produce small amounts of all three types of hormone, as well as steroid hormones that do other things (such as altering mineral and water balance in the body).

The secretion of all these hormones, whether from ovary, testis or adrenal glands, is under the control of the pituitary gland, which is situated at the base of the brain. It produces hormones that in turn control the output of most of the other endocrine glands. Thus, it is the pituitary that 'manages' the complex hormonal events of the menstrual cycle. Because it is so close to the brain, the pituitary is highly susceptible to influences from it: in this way, stress can affect the menstrual cycle. Women – especially younger women – starting a new job, or travelling to a strange place, often find that their periods become delayed.

HORMONES BEFORE BIRTH: HOW WE BECOME WOMEN OR MEN

Very early on in embryonic life (before the seventh week after conception in humans), hormones have a role in differentiating the sex of the foetus. At this age, foetuses are not sexually differentiated. If the embryo is genetically female (that is, if it has two X chromosomes: see Chapter 10), then an ovary develops, followed by fallopian tubes and uterus inside, and clitoris, vagina and vulva outside. If, however, the embryo is male (that is, if it has an X and a Y chromosome), then testes develop, which produce androgens, which proceed to masculinise the embryo. Inside, the tubes that would become fallopian tubes in a female embryo disappear, and a different set of tubes develops, eventually to become the vas deferens. Outside, the originally neutral tissue develops, not into clitoris and vulva, but into penis and scrotum. All of these changes occur under the influence of the hormones from the embryonic testes, which divert development into the 'male' pattern. The female pattern, therefore, is the basic one: in the *absence* of any androgens, the female pattern will develop. It requires the active intervention of 'male' hormones to make a male. In animals that produce live young such as ourselves, some of the mother's hormones get into the foetal circulation by way of the placenta. If sexual differentiation were dependent on the presence of female hormones, all embryos would be exposed to them through the mother's circulation, and all would be female. By making sexual differentiation depend on the presence or absence of androgens from the foetal testes, nature has ensured that there are two sexes. The advantage of having two sexes lies in *sexual* reproduction, rather than, say, reproduction by division. Sexual reproduction helps to maintain genetic diversity, because each individual inherits genetic material, and characteristics, from *both* parents. Nearly all species of animals reproduce sexually, although some can also reproduce asexually (for example, by dividing into two).

In other mammals, the same presence or absence of androgens can influence the brain so that the adult animal tends to show predominantly 'male' or 'female' behaviour accordingly. Thus a genetically female rat can be given androgens as soon as she is born, and in adulthood she will behave sexually more like a male does (for example, she will be more aggressive than most females, and will not become receptive to the advances of males). As the brain effects occur slightly later than the physical effects, the hormones can be given *without* affecting her external genitalia, so that she looks like, but does not behave like, her untreated sisters. This type of experiment can tell us how behaviour becomes differentiated in animals, but it tells us little about humans. There is, in fact, little evidence of any strong effects of hormones on brain and behaviour before birth in humans (although there *may* be some slight effects which are at present not fully known). Our behaviour patterns are considerably more dependent on complex cultural learning than those of a rat. For example, removal of a rat's ovaries will make her uninterested in mating, which is certainly not true of women, who continue to be sexually active long after their ovaries have been removed (or have become quiescent, as in a natural menopause).

THE MENSTRUAL CYCLE

The female cycle and its control are immensely complex. Having said that, we will try to explain – simply – the events occurring in each stage of the cycle. Table 7 charts the events in the uterus, ovaries and vagina throughout the menstrual cycle. To simplify the hormonal events considerably, we can say that: (1) oestrogen dominates the first phase, its levels increasing slowly from the start of menstruation until it reaches a peak just before ovulation; (2) both progesterone and oestrogen (progesterone is the progestin normally produced by the ovary) are present in the second half of the cycle; and (3) blood levels of both hormones fall a few days before the next

Table 7: The Events of the Menstrual Cycle

Uterus (Womb)	What is happening to the: Ovaries	Vagina	How some women say they feel: Pains	Moods
Stage 1: Menstruation The lining starts to break down and is shed.	A new egg (ovum) begins to grow in one of the ovaries.	Vaginal secretions change from thick and white just before to scanty by the end.	Often cramps and/or low back pain, especially at the beginning.	Tiredness to start with. May feel more sexy.
Stage 2: End of menstruation to about day 12* (follicular phase) A new lining starts to form, but, as yet, it is not very thickly lined with blood vessels.	The new ovum becomes a 'graafian follicle' in which the ovum is surrounded by a sac of fluid.	Vaginal mucus is quite thin, and is alkaline. If allowed to dry on glass, it forms a 'fern' pattern (see Figure 4)		Often feel energetic, able to do things; outward going.
Stage 3: Ovulation, about day 13–15* A new lining has developed during stage 2, ready to receive an ovum if it is fertilised.	Under influence of pituitary gland, ovum bursts out of follicle. Small amount of bleeding may accompany ovum release from ovary.	Mucus becomes very thin and stretchy. Larger quantity than usual. Readily shows fern pattern on glass.	A few women feel slight abdominal pain as the follicle ruptures.	Peak of energy. Some women feel sexier.
Stage 4: The luteal phase: from day 15 to about day 25* The glands and blood-vessels become more tightly folded to receive and nourish a fertilised ovum if there is one.	Tissue left behind in ovary produces hormones which can maintain a pregnancy if necessary until the placenta is big enough to take over. These hormones cause the changes in the uterus.	Mucus is thick and white, containing cells from the vagina.		More inward-looking and reflective. Sometimes more tired. Time of lowered sex drive in some women.
Stage 5: From day 25* until the start of the next period Glands and blood-vessels reach maximum folding. As hormone levels beginning to fall, the blood-vessels begin to constrict, ready for bleeding.	Hormone levels start to decline from the ovary. It is this which eventually brings about the bleeding.	As in previous stage, though may become thicker.	May be some pain in the uterus, general body pains, nausea.	Many women report marked behaviour changes in this stage: irritability, depression, anxiety, etc.

*Dating the cycle starts from the *first* day of menstrual bleeding.

period is due. The decline in hormone levels induces the bleeding. The Pill works by mimicking phase (2). Nearly all pills contain an oestrogen and a progestin. It works by 'fooling' the brain that the body is in phase (2), and has *already* ovulated. Hence, the pituitary does not send out its hormones, which would otherwise induce ovulation. In a normal menstrual cycle, the oestrogen plus progesterone, which characterises the period after ovulation, work in the same way. They effectively inhibit the pituitary from sending out any more signals to ovulate.

The chart is based on the *average* cycle length of twenty-eight days, although we know that few women actually have cycles of that length. However, we have included in the chart descriptions of changes that are not usually described in detail, as many women like to know what is happening to their bodies during the cycle, and to find out, for example, when they ovulate.

In women who are ovulating regularly, there is a slight rise in basal body temperature (that is, the body temperature when you are being inactive before you get up in the morning). For women who happen to have access to a microscope, or who wish to buy one, changes in the cells of both the vagina and, to a lesser extent, the inside of the mouth can be observed. During the oestrogen-dominated phase (up till mid-cycle), the cells are predominantly cornified (see Figure 3), especially at ovulation itself. From then until menstruation the cells are mainly leucocytes, which are considerably smaller. You may be able to observe these changes in your own cells, but be warned: it takes a bit of practice. You may need to watch and record carefully, perhaps even draw, the cells you see each day that you do your observations, but if you look carefully you will observe the changes. You only need to do this every few days, except around mid-cycle, when the changes may occur quite rapidly. It is best to do it daily around mid-cycle.

The appearance of vaginal and cervical mucus smeared on to glass changes with the cycle too (see Figure 4). You can see the changes in the consistency of the mucus, as well as in the colour of the cervix (the neck of the womb) if you have a speculum and try self examination. (For more information about self-examination see *The New Women's Health Handbook*, edited by Nancy MacKeith, and published by Virago.) If you are on the Pill none of these changes will be very noticeable, as the Pill imposes more constant hormone levels on your body.

Different women experience different feelings during the premenstrual period. Some become irritable, some feel pains, some

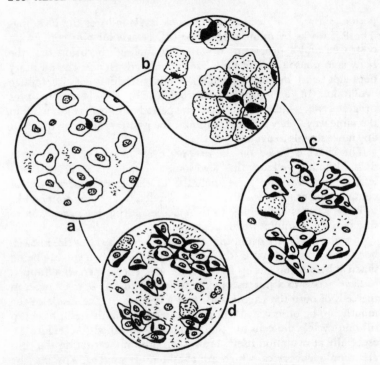

Figure Three: *The appearance of vaginal cells under a high-powered microscope.*

These are cells taken from the vaginal walls at different stages of the menstrual cycle. (a) shows cells taken shortly after menstruation has ended; (b) shows cells at ovulation; (c) shows cells shortly after ovulation; and (d) shows cells a few days before menstruation.

find their appetite is increased, and so on. These feelings vary a lot between women, so we have only included a few in the table. Even the physical feelings seem to vary between women, and between cycles: you may find that your breasts do not become sore at every cycle, for instance. No one really knows why this variation occurs, although it is probable that you don't produce *exactly* the same quantities of hormones each cycle and that you feel the soreness only under certain hormone conditions.

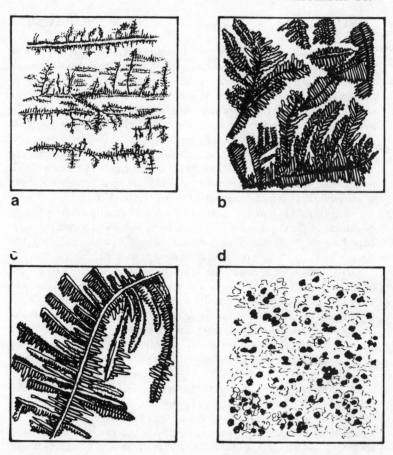

Figure Four: *The appearance of vaginal mucus under a low-powered microscope.*

The appearance of vaginal mucus changes with the menstrual cycle. (a) shows the appearance of mucus from early in the menstrual cycle, shortly after menstruation has finished; (b) shows the appearance of mucus just before ovulation, approximately ten days after the previous period started; (c) shows the appearance of mucus at the time of ovulation; (d) shows the appearance of mucus after ovulation, during the second half of the cycle.

WHAT HAPPENS DURING THE MENOPAUSE?

It used to be thought that the menopause occurred as ova 'ran out' in the ovary, but this is no longer thought to be the case. We now know that there are in fact many mature ova left in the ovaries of menopausal women, but that somehow they do not respond to the usual signal from the pituitary which previously had resulted in ovulation. From studies of mice and rats we know that, if 'old' ovaries are transplanted into young animals, they behave cyclically, producing both ova and hormones. Similarly, if 'young' ovaries are transplanted into old animals, they *also* behave normally. This means that the pituitary gland of the old animals (which normally 'manages' the hormone output of other glands) is still capable of working normally. So the ovary *can* work normally given a young pituitary, and the pituitary *can* work normally given a young ovary. Somehow or other, the link between them gets broken during the menopause. Quite how remains a mystery at present.

As this link breaks down, the levels of hormones produced by the ovary decline. This seems to be the result of decreasing sensitivity of the ovary to instructions from the pituitary. It is the declining levels of oestrogens that are involved in many of the symptoms of the menopause, such as hot flushes and insomnia. Oestrogens never entirely disappear, however: small quantities are still produced by the adrenal glands throughout a woman's life.

Studies of ageing animals have also suggested that the uterus is affected by ageing too. It becomes less sensitive to hormones, and its cells no longer change cyclically. It also tends to reject any fertilised ova that may be present, rather than allowing them to implant, as would happen in younger animals. Perhaps it is just as well, as older animals produce more abnormal ova than do younger ones. It is well known that the incidence of congenital defects is much higher in offspring born to older mothers, and increases with maternal age. The most likely function of the menopause, then, is to prevent the possibility of defective embryos.

An older woman may experience many physical changes that are the result of the menopause, such as drying of the vagina, hot flushes, bouts of dizziness and so on. She may also experience changes that are more due to general ageing, such as tiredness and a tendency to bone brittleness (osteoporosis). Although these may be influenced by the declining hormone levels, they are really just ageing processes. Oestrogen levels decline by about 80 per cent after the menopause, and during it they seem to vary considerably.

Oestrogens normally tend to inhibit the pituitary from producing certain of its hormones (those that normally affect the reproductive system). Consequently, after oestrogen levels have fallen in the menopause the pituitary hormones are increased, but they no longer have any effect on the ovaries.

BIRTH

About a century ago, physiologists thought that it was the mother's body that initiated birth, and that the foetus was simply a passive traveller from its rather cushioned existence in the womb to the harsher world outside. Although we still do not know enough about it, we do know that the foetus in fact plays a very decisive role in determining when it is going to be born. Much of what is now known has been discovered from experiments with sheep, though there is no reason to believe that human beings are significantly different.

At the end of the period of gestation, the foetal brain initiates a train of events that culminate in birth. Activity in part of the foetus's brain (the hypothalamus) causes the pituitary gland to secrete more hormones, including the one that stimulates the adrenal glands, which in response produce their own hormones. Adrenal hormones (glucocorticoids) are conveyed by the blood to the placenta, where they exert two effects. First, they cause the amount of progesterone produced by the placenta to be reduced. The effect of this is to increase the activity of the muscles of the womb (normally, progesterone inhibits their activity), which helps in the early stages of labour. It was once thought that the withdrawal of progesterone was of primary importance, but it is now known that a second factor – prostaglandins – is more important. As the foetal adrenals produce more of their hormones, they stimulate the mother's uterus to produce a substance called prostaglandin F_2a (PGF_2a). This happens about twenty-four hours before labour. This substance is very powerful in stimulating the womb to contract, and has been used by doctors to induce labour, or to induce abortions between three and six months of pregnancy.

In the last stages of labour, when more forceful contractions are needed than have been induced by the mother's prostaglandins, her own pituitary gland starts to produce another hormone called oxytocin (which is also used medically in birth inductions). This hormone induces even stronger contractions from the uterus, which expel the baby: it also acts on the breast and causes ejection of milk (this is why breastfeeding is recommended to help to get the womb back to its normal size: as the baby suckles, oxytocin causes the milk to be ejected into its mouth, and it also affects contractions of the uterus). All these stages are summarised in the chart.

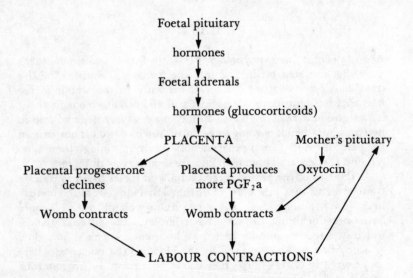

Notes and References

INTRODUCTION

1. Critiques of the existing form of science from a marxist perspective have appeared in several places. For example, see issues of the *Radical Science Journal*, and *Science for People*. Also see *The Political Economy of Science*, and *The Radicalisation of Science*, both edited by Rose H., and Rose, S., London, Macmillan, 1976.

2. This notion varies somewhat: sometimes it is stated as though 'aggression' (the definition of which is not usually stated) is a product of (male) hormones; alternatively, it might be a product of the Y chromosome (which only biological males have among mammals), or a product of our evolutionary heritage. It is articulated by such popular writers as Robert Ardrey, Lionel Tiger and Robin Fox.

3. See, for example, Kamin, L., *Science, Politics and IQ*, Harmondsworth, Penguin, 1977, and Lawler, J. M., *IQ, Heritability and Racism: A Marxist Critique of Jensenism*, London, Lawrence & Wishart, 1978.

4. See, for example, Easlea, B., *Liberation and the Aims of Science*, London, Chatto & Windus, 1974, pp.191–203: and Rose, H. and Rose, S., *Science and Society*, Harmondsworth, Penguin, 1969, pp.128–9.

5. Rose, H. and Rose, S., 'The Incorporation of Science', in Rose, H. and Rose, S. (eds), *The Political Economy of Science*, London, Macmillan, 1976, p.14.

6. Haslemere Group pamphlet, *Who Needs the Drug Companies?*, London, Women in Print, 1976.

7. See the sources listed under n.1 above.

8. Rose and Rose, 'The Incorporation of Science', op. cit. pp.14–15.

9. 'Uncontrollable' here can mean 'not easily controlled by the state and its institutions'. For an analysis of the methods employed by the state in Britain to control certain groups of people (e.g. those involved in particular political groups), see Ackroyd, C. *et al.*, *The Technology of Political Control*, Harmondsworth, Penguin, 1977, part 4.

10. On the naive assumption that we can identify – and therefore remove – bits of the brain that determine specific behaviours in people. It was precisely on this assumption that this suggestion was made. Removal of specific brain areas (such as parts of the structure called the amygdala) in other mammals might make them less likely to bite others: what this means for people is anyone's guess.

11. See Rose, H. and Rose, S., 'The Politics of Neurobiology: Biologism in the Service of the State', in Rose and Rose, *The Political Economy of Science* op. cit.; and Ackroyd *et al.*, op.cit. pp.275–80.

12. Antoinette Brown Blackwell, 'Sex and Evolution' (1875), in Rossi, A. S. (ed.) *The Feminist Papers*, New York, Bantam, 1973.

13. Ehrenreich, B. and English, D., *Complaints and Disorders: The Sexual Politics of Sickness*, London, Compendium, 1974, p.27.

14. ibid., pp.45–62.

15. Moebius, P.; cited by Ehrenreich and English, op.cit., p.28.

16. Webb, B., *My Apprenticeship*, Harmondsworth, Penguin, 1971.

17. See Kamin, op. cit., especially Chapter 1.

18. Wright, Sir E. A., *The Unexpurgated Case Against Woman Suffrage*, London, Constable, 1913, p.36.

19. R. R. Coleman; cited by Haller, J. S. and Haller, R. M., *The Physician and Sexuality in Victorian America*, Urbana, University of Illinois Press, 1974, p.39.

20. Quoted in Wood, C. and Suitters, B., *The Fight for Acceptance: A History of Contraception*, London, Medical and Technical Publishers, 1970, p.143.

21. Feminist work on science includes: Hubbard, R., Henifin, M. S. and Fried, B., *Women Look at Biology Looking at Women: A Collection of Feminist Critiques*, Boston, G. K. Hall and Co., 1979. There was also a special issue of the journal *Signs* dedicated to women and science in Autumn 1978. There have also been special issues of *Science for People* and *Undercurrents* magazines devoted to consideration of women and science.

22. Alam, A., 'Science and Imperialism', *Race and Class*, 19 (1978), p.239. The idea is discussed at greater length by Gorz, A., 'On the Class Character of Science and Scientists', in Rose and Rose, *The Political Economy of Science*, op.cit.

23. This is not to say that certain features of other branches of science, such as physics or chemistry, do not lend themselves to oppression. However, it is the biological sciences that contribute uniquely to *women's* oppression, simply because we are defined so often in terms of our biology and the limitations that that supposedly places upon us. Thus, we are not so much evading the 'hard' sciences of physics and chemistry (which, of course, are heavily dominated by men in terms of numbers working within those fields), as concentrating on those aspects of science that are directly relevant to the ideology of women's oppression.

24. Sociobiology is the study of animal societies, and their evolutionary development. See Chapter 3.

25. Bowlby, J., *Maternal Care and Mental Health*, Geneva, World Health Organisation, 1951.

26. See, for example, Rutter, M., *Maternal Deprivation Reassessed*, Harmondsworth, Penguin, 1972.

27. ibid., pp.120–8.

28. There has been a suggestion (with which we do not entirely agree) that the terrible ways in which our society has developed – what it is doing to itself – are ultimately the product of a woman-dominated pattern of childrearing. See Dinnerstein, D., *The Rocking of the Cradle and the Ruling of the World*, London, Souvenir Press, 1978.

29. Rutter, op.cit., pp.105–14.

30. Graham, H. and McKee, L., 'Ideologies of Motherhood and Medicine on Radio and Television', manuscript of paper presented at the British Sociological Association meeting, University of Sussex, 1978.

31. Ehrenreich, B. and English, D., *For Her Own Good: 150 Years of the Experts' Advice To Women*, New York, Anchor Press, 1978, pp.128–31.

32. ibid., p.128.

33. ibid., p.129.

34. ibid., pp.141–5.

35. ibid., pp.146–8.

36. ibid., p.141.

37. Different types of science education for different social groups were not only advocated for the two sexes. During the nineteenth century the Mechanics Institutes were established, aiming at providing some scientific education for the working classes, in the belief that 'a regimen of scientific education for certain members of the working class would render them, and their class as a whole, more docile, less troublesome, and more accepting of the emerging structure of industrial society'. See Shapin, S. and Barnes, B., 'Science, Nature and Control: Interpreting Mechanics' Institutes', in Dale, R., Esland, G. and MacDonald, M. (eds), *Schooling and Capitalism: A Sociological Reader*, London, Routledge & Kegan Paul/Open University Press, 1976.

38. In abbreviating so much, we are distorting these descriptions of the process of science to some extent. Our intent was to sketch out approximately what scientists do, rather than to involve ourselves in analysing the now extensive literature on the philosophy and sociology of science. Interested readers are referred to Lakatos, I., and Musgrave, A., *Criticism and the Growth of Knowledge*, Cambridge University Press, 1970; Kuhn, T. S., *The Structure of Scientific Revolutions*, 2nd edn, Chicago University Press, 1970; and several papers in Barnes, B. (ed.), *Sociology of Science*, Harmondsworth, Penguin, 1972.

39. For an outline of the rise of science, see, for example, Easlea, op.cit., and Rose and Rose, *Science and Society*, op.cit.

40. Rose and Rose, *Science and Society*, pp.16–36. Also see Ben-David, J.

and Zloczower, A., 'The Growth of Institutionalised Science in Germany', in Barnes, op.cit.

41. Human dominion over the world has indeed increased through scientific progress. We do not, however, intend to convey an idea that this is inherently bad, and we are certainly not advocating a 'back to nature' idea which rejects all scientific knowledge out of hand. Our extended dominion, of course, includes that over disease (although not all peoples of the world benefit equally from this knowledge). The idea with which we are concerned here is primarily the attitude that we can dominate, rather than interact with, nature.

42. The result of this view of the native American Indians is documented by Dee Brown, *Bury My Heart At Wounded Knee: An Indian History of the American West*, London, Barrie & Jenkins, 1971.

43. McLuhan, T. C., *Touch The Earth: A Self-Portrait of Indian Existence*, London, Sphere Books, 1973, p.15.

44. See Ackroyd, C. *et al.*, op.cit., Part IV, pp.229–84.

45. The operation of science for profit, even at the expense of people, is brought home in Kay Weiss's article, 'Vaginal Cancer: An Iatrogenic Disease', *International Journal of Health Services*, 5(2) (1975), pp.235–51. She discusses the reluctance of the US Food and Drug Administration to curb the use of diethylstilboestrol, after evidence that it caused vaginal cancer in the daughters of those women to whom it was given (for threatened miscarriage). Federal agencies are reluctant, she points out, to infringe on the rights of private corporate interests – a position summed up by the FDA's chief counsel: 'Industry is likely to challenge in the courts any FDA action where the net adverse economic impact exceeds the legal fees involved.' Meanwhile, the drug was – and still is – in use, given to thousands of women as a 'morning-after' pill. Most of these women were told neither of its dangers, nor of the fact that the drug was still on trial. For a more detailed, but immensely readable, critique of this unquestioning use of hormones, see Seaman, B. and Seaman, G., *Women and The Crisis in Sex Hormones*, Hassocks, Harvester Press, 1978.

46. Benefit, that is, in terms of the possibility of avoiding endless pregnancies. We recognise that the use of the Pill is accompanied by certain risks to health and life, and that it is often used under conditions which probably do *not* help women (such as clinical trials on ill-informed and illiterate women in developing countries).

47. With the welcome exceptions listed under n.21 above.

CHAPTER 1
SCIENCE EDUCATION: DID SHE DROP OUT OR WAS SHE PUSHED?

1. *Statistics of Education*, London, HMSO, 1975.

2. ibid.

3. In Britain, few managing directors have had a science training. This is not true in EEC countries. But there are still no women.

4. *Sample Census*, 1966, Scientific and Technological Qualifications, Table 6.

5. *Statistics of Education*, op.cit.

6. Kelly, A., 'A Discouraging Process: How Women are Eased Out of Science', paper presented at the conference on 'Girls and Science Education', Chelsea College, London, March 1975.

7. *Times Educational Supplement*, 26.5.1978, p.19.

8. *Equal Opportunities for Men and Women*, London, Department of Education/HMSO, 1973.

9. *Statistics of Education*, op.cit.

10. *Times Educational Supplement*, op.cit.

11. Fairweather, H., 'Sex Differences in Cognitive Tests', *Cognition*, 4. (1976) pp.231–75.

12. See for example, Rosenthal and Jacobson, *Pygmalion in the Classroom: Teacher Expectations and Pupil's Intellectual Development*, New York, Holt, Rinehart, and Winston, 1968.

13. *1906 Report of the Board of Education*, in van der Eyken, Willem (ed.), *Education, the Child and Society*, Harmondsworth, Penguin, 1973, p.126.

14. Regulations for Secondary Schools (1909), in *Oxford Review of Education*, 3, (I) (1977), pp.26–8.

15. *1906 Report*, op.cit., p.138.

16. Newsom, *Half Our Future*, London, Central Advisory Council for Education/HMSO, 1963.

17. ibid.

18. Bantock, G. H., *Towards a Theory of Popular Education*, in Hopper, Richard (ed.), *The Curriculum*, Oxford University Press, 1971, p.263.

19. Ashton, P. *et al*, *The Aims of Primary Education: A Study of Teacher's Opinions*, London, Macmillan, 1975.

20. 'From about 1940–1950 girls usually did better than boys at the intelligence and arithmetic tests in 11-year selection examinations, as well as at English, and different borderlines had to be drawn for entry to grammar schools.' (Vernon, P. E., *Intelligence and Attainment*, University of London Press, pp.170–1.)

In 1957 a report on the allocation of grammar school places made the following recommendations:

> the tendency for girls to obtain higher test scores than boys during the final year of the primary school course is probably associated with the tendency for girls to mature earlier than boys, and as boys at a later stage reach puberty and enter the phase of rapid development, they are likely to recover their lost ground. . . . If the level of performance at the age of transfer represented enduring differences then it would be logical to award a larger proportion of grammar school places to girls than to boys. If on the other hand the differences are only temporary, such a procedure would be unjust.
>
> It follows that if in an area equal numbers of grammar school places are made available to boys and girls, the pass marks will necessarily have to be fixed at different levels. This inevitably involves the situation in which a boy and a girl, possibly a brother and a sister, might produce identical levels of performance in the authority's examination but could not both be awarded a grammar school place. If alternatively the pass marks are made equivalent for the two sexes, the numbers of girls admitted to grammar schools in most areas will substantially exceed the number of boys.
>
> In view of the fact, therefore, that these differences exist at the age of 11 and there is considerable uncertainty as to when and to what extent they eventually disappear, the most satisfactory course for an authority to adopt would be to treat boys and girls separately for the purpose of allocation to secondary schools. [Yates, A. and Pidgeon, D. A., *Admission to Grammar Schools, 3rd Interim Report on the Allocation of Primary School-leavers to Courses of Secondary Education*, London, Hazell: Watson & Viney, 1957, pp.167–8]

This seems to be a roundabout and obscure way of saying that, since boys are late developers, authorities should set a lower pass mark for them in order to ensure equal numbers of boys and girls in the grammar schools. In 1961, for example, there were 352,514 boys and 344,163 girls at grammar school.

21. DES circular, 'Differences for Boys and Girls', *Education Survey*, 21.

22. Byrne, E. (ed.), 'Training and Equal Opportunities', unpublished paper, Deputy Chief Education Officer, City of Lincoln, 1973.

23. DES circular, op.cit.

24. Dale, R. R.: *Mixed or Single Sex Schools*, 2 vols, London, Routledge & Kegan Paul, 1969–71.

25. DES circular, op.cit.

26. Before allowances are made for various factors, such as rural schools.

27. Phillips, C., *Changes in Subject Choice at School and University*, LSE Research Monographs, University of London Press, 1969.

28. Phillips, op.cit.

29. *Brighton and Hove Women's Liberation Newsletter*, July 1978.

30. Data from a questionnaire survey administered by A. F. McPherson to all new entrants to Edinburgh University in 1972.

31. *Sample Census*, op.cit., Table 6.

32. 'Many collaborations are set up informally and results circulate well before publication but women do not fit easily into this system. Often an older man takes a young colleague under his wing and helps him meet the right people and land the right jobs but this is not so easily done for a woman.' (Kelly, op.cit.)

33. *New Scientist*, 3.8.1978.

34. Ashton, B. G. and Meridith, H. M., 'The Attitudes of Sixth-Formers', in Hinton, K. (ed.), *Women and Science: Science in a Social Context*, SISCON, pp.42–3.

35. Practical knowledge is not in itself 'low-status'. Medicine is an example of an 'applied' field that leads to a high-status occupation. But the way to get into medical school is to do pure science A levels, not woodwork.

36. At present, money and resources are concentrated on educating a few experts and specialists for the top jobs, rather than on developing everybody's potential to the full. About four times as much is spent on a university student as on a primary school child; five times as much is spent on a university student as on an FE student.

37. Our view of EOC Review of TOPS, *Women in Manual Trades Newsletter*, February 1979.

38. Woman scientist, quoted by Liz Manton, '. . . and here begins our alienation from science', *Undercurrents*, August–September 1978, pp.14–15.

39. ibid.

40. Watson, D. and Crick F., *The Double Helix*, Harmondsworth, Penguin, 1970.

41. Quotations from Manton, op. cit.

CHAPTER 2
PSYCHOLOGICAL SEX DIFFERENCES

1. Freud, S., 'Femininity' (1933), in *New Introductory Lectures on Psychoanalysis*, Harmondsworth, Penguin, 1973, p.149.

2. See, for instance, de Beauvoir, S., *The Second Sex* (1949); Friedan, B., *The Feminine Mystique* (1963); Millett, K., *Sexual Politics* (1969); Figes, E., *Patriarchal Attitudes* (1970); Firestone, S., *The Dialectic of Sex* (1970); and Mitchell, J., *Psychoanalysis and Feminism* (1974). For a recent review of these and other feminist critiques of Freud see Sayers, J., 'Anatomy is Destiny: Variations on a Theme', *Women's Studies International Quarterly* 2 (1979), 19–32.

3. Brown, D., 'Sex-role Preference in Young Children', *Psychological Monographs* 70 (14) (1956), pp.1–19.

4. Given that this and other psychological tests symbolically endorse, through their scoring systems, the low value placed by society on feminine responses, it is ironic that psychologists should also urge women to place value on 'their particular talents and skills'. Hutt, C., *Males and Females*, Harmondsworth, Penguin, 1972, p.138.

5. Brown, D., 'Masculinity–Femininity Development in Children', *Journal of Consulting Psychology* 21 (1957), p.202.

6. Terman, L. M., and Miles, C. C., *Sex and Personality*, New York, Russell and Russell, 1936.

7. Although the main drift of Terman and Miles's book is to imply such an equation, their actual findings indicate that this equation is not universally true. Similarly, although attempts have been made to correlate high masculinity scores on the M–F scale of the Minnesota Multiphasic Personality Inventory (the MMPI) with lesbianism, such attempts have failed. See, for instance, Dahlstrom, W. G., and Welsh, G. C., *An MMPI Handbook*, Minneapolis, University of Minnesota Press, 1960.

8. Bem, S. L., 'The Measurement of Psychological Androgyny', *Journal of Consulting Clinical Psychology* 42 (1974), pp.155–62.

9. Freud, op.cit.

10. Freud bemoaned the situation, for instance, in which a girl might resolve her early 'penis envy' by giving up 'her phallic activity and therewith her sexuality in general and a considerable part of her masculine proclivities in other fields'. See his 1931 paper, 'Female Sexuality' in *Collected Papers*, Vol. V, London, Hogarth, 1950, p.257.

11. Bem, S. L., 'Sex Role Adaptability: One Consequence of Psychological Androgyny', *Journal of Personality and Social Psychology* 31 (1975), 634–43; and Bem, S. L., Martyna, W., and Watson, C., 'Sex Typing and Androgyny: Further Explorations of the Expressive Domain', *Journal of Personality and Social Psychology*, 34 (1976), pp.1016–23.

12. For a general review of the ways in which M–F tests have wrongly assumed non-overlap of masculine and feminine traits, and have trivialised the study of female and male psychology, see Constantinople, A., 'Masculinity-femininity: An Exception to a Famous Dictum?' *Psychological Bulletin*, 80 (1973), pp.389–407.

13. Affirmative responses to any of these questions receive a minus (i.e. 'feminine') score, while negative responses receive a plus (i.e. 'masculine') score in Terman and Miles's scoring system.

14. Intelligence tests are quite regularly used to decide whether a child should be placed in a special school for the educationally subnormal. They are also still used in some counties to select children for different forms of secondary education.

15. Joanna Ryan makes this point in Richardson, K., and Spears, D. (eds), *Race, Culture and Intelligence*, Harmondsworth, Penguin, 1972.

16. This is a clear implication of Hutt's book, op. cit.

17. Kagan, J., 'Acquisition and Significance of Sex Typing and Sex Role Identity', in Hoffman, M. L. and Hoffman, L. W., (eds), *Review of Child Development Research*, Vol. I, New York, Russell Sage, 1964, p.146.

18. Williams, J. E., Bennett, S. M. and Best, D. L., 'Awareness and Expression of Sex Stereotypes in Young Children', *Devel. Psychol.* 11 (1975), pp.635–42.

19. Rosenkrantz, P., Vogel, S., Bee, H., and Broverman, D., 'Sex-role Stereotypes and Self-concepts in College Students', *J. Consult. Clin. Psychol.* 32 (1968), pp.287–95.

20. Goldberg, S., and Lewis, M., 'Play Behaviour in the Year-Old Infant: Early Sex Differences', in Bardwick, J. M. (ed.), *Readings on the Psychology of Women*, New York, Harper, 1972.

21. Maccoby, E. E., and Jacklin, C. N., *The Psychology of Sex Differences*, Stanford University Press, 1975.

22. Barbara Lloyd suggests that the non-publication of valid findings of no sex difference is due to statistical considerations. Whatever the reason, the result is, as she points out, that 'we have little or no evidence of behaviours in which no differences in the performance of men and women have been found'. See Lloyd, B. and Archer, J. (eds), *Exploring Sex Differences*, London, Academic Press, 1976, p.9. Maccoby and Jacklin, op. cit., p.4, also comment on the failure of psychologists to report findings of no sex difference.

23. It is a general tendency in science that negative results are not considered worth publishing. Even if the author thinks so, the publishing editor probably will not! [Eds]

24. Hutt, op. cit.

25. Garai, J. E. and Scheinfeld, A., 'Sex Differences in Mental and Behavioural Traits', *Genetic Psychol. Monogr.* 77 (1968), p.270.

26. See Maccoby and Jacklin, op. cit.

27. Nash, J., *Developmental Psychology: A Psychobiological Approach*, Englewood Cliffs, Prentice-Hall, 1970, p.355.

28. ibid.

29. Rosenkrantz *et al.*, op. cit.

30. Maccoby and Jacklin, op. cit.

31. Fairweather, H., 'Sex Differences in Cognition', *Cognition*, 4 (1976), pp.231–80.

32. Maccoby and Jacklin, op. cit., p.355.

33. Strathern, M., 'An Anthropological Perspective', in Lloyd and Archer, op. cit., p.53.

34. Mill, J. S., *The Subjection of Women*, London, Longmans, Green, Reader, and Dyer, 1869.

35. Bagehot, W., 'Biology and "Women's Rights"', *Popular Science Monthly*, 14 (1879), p.208.

36. Nash, op. cit., p.204.

37. Social Darwinists were social theorists who tried to apply Darwin's theory of evolution to human society in an attempt to justify the then current social order. For a review of the arguments put forward by them against changing the status of women, see Fee, E., 'Science and the Woman Problem: Historical Perspectives', in Teitelbaum, M. S. (ed.), *Sex Differences*, New York, Anchor Books, 1976.

38. Hutt, op. cit., p.136.

39. ibid., p.138.

40. Nilsson, A. (1970); cited in Breen, D., *The Birth of a First Child*, London, Tavistock, 1975.

41. Hutt, op. cit., p.133

42. ibid., p.108.

43. Maccoby and Jacklin, op. cit.

44. Gray, J. A., and Buffery, A. W. H., 'Sex Differences in Emotional and Cognitive Behaviour in Mammals including Man: Adaptive and Neural Bases', *Acta Psychologica*, 35 (1971), pp.89–111.

45. Hutt, op. cit., p.97.

46. Hutt does not comment on the fact that the preponderance of women in these occupations holds only at certain times, and even then only in some societies. In our own society, for instance, typists used to be men rather than women; and in other societies today, needlework, for example, is men's, not women's, work.

47. Thompson, H., *The Mental Traits of Sex*, University of Chicago Press, 1903.

48. Swinburne, J., 'Feminine Mind Worship', *Westminster Review*, 158 (1902), p.189.

49. Maccoby and Jacklin, op. cit., p.130.

50. See, for instance, Newcombe, F. and Ratcliff, G., 'The Female Brain: A Neuropsychological Viewpoint', in Ardener, S. (ed.), *Defining Females: the Nature of Women in Society*, London, Croom Helm, 1978.

51. Coltheart, M., 'Sex and Learning Differences', *New Behaviour*, 1.5.1975, pp.54–7.

52. Nash, op. cit., p.210.

53. Wilby, P., 'Why Our Education System is Unfair to Women', *Sunday Times* (colour supplement), 12.3.1978, p.22.

54. There is another view – see Chapter 1. [Eds.]

55. Hutt, op. cit., p.18.

56. See, for instance, Freud, S., 'The Passing of the Oedipus Complex', in *Collected Papers*, Vol. II, London, Hogarth, 1924; and Freud, S. (1925). 'Some Psychological Consequences of the Anatomical Distinction between the Sexes', in *Collected Papers*, Vol. V, London, Hogarth, 1950.

57. Money, J., Hampson, J. G. and Hampson, J. L., 'Hermaphroditism: Recommendations Concerning Assignment of Sex, Change of Sex, and Psychologic Management', *Bulletin of Johns Hopkins Hospital*, 97 (1955), pp.284–300.

58. Hormones and chromosomes are indicative of biological sex in so far as, normally, women have lower levels of androgens relative to the levels of the other two sex hormones – oestrogen and progestin – than do men, and their sex chromosomes consist of a pair of X chromosomes, whereas the sex chromosomes of men consist of an X chromosome paired with a smaller Y chromosome.

59. See, for instance, Wiechel, L., 'Sex-role Perception as a Barrier to Cooperation: Assessment Techniques and Programmes of Educational Influences', *Didakometry and Sociometry*, 5 (1973), pp.1–8.

60. It has been suggested, for instance, that women's lack of occupational achievement is due to their 'fear of success', particularly to fear of success in those occupations that are stereotyped as 'masculine', or in which men are the more numerous sex. For a review of the research on this subject see O'Leary, V. E., *Toward Understanding Women*, Monterey, Brooks/Cole, 1977, Chapter 5.

61. Almquist, E., M. and Angrist, S. S., 'Role Model Influences on Women's Career Aspirations' (1971), in Theodore, A. (ed.), *The Professional Woman*, Cambridge, Schenkman, 1971.

62. Thompson Woolley, H., 'A Review of the Recent Literature on the Psychology of Sex', *Psychological Bulletin*, 7 (1910), p.340.

63. Freud, S. (1933), op. cit., pp.147–8.

64. ibid., p.148.

65. See, for instance, Freud's (1925) paper (op. cit.), in which his thoughts about female psychology lead him to regard as problematic the 'prehistory of the Oedipus complex', and to refer to this early period of development as consisting of 'dark regions where there are as yet no sign-posts' (p.186).

66. See, for instance, Terman, L. M., and Tyler, L. E., 'Psychological Sex Differences', in Carmichael, L. (ed.), *Manual of Child Psychology*, 2nd edn, New York, Wiley, 1954.

67. See, for instance, Kohlberg, L., 'A Cognitive-developmental Analysis of Children's Sex-role Concepts and Attitudes', in Maccoby, E. E. (ed.), *The Development of Sex Differences*, Stanford University Press, 1966.

68. See, for instance, Mischel, W., 'Sex-typing and Socialisation', in

Mussen, P. H. (ed.), *Carmichael's Manual of Child Psychology*, 3rd edn, New York, Wiley, 1970.

69. This point is made, for instance, by Mussen, P. H., 'Early Sex-role Development', in Goslin, D. A. (ed.), *Handbook of Socialisation Theory and Research*, Chicago, Rand McNally, 1969; and by Maccoby and Jacklin, op. cit.

CHAPTER 3
SOCIOBIOLOGY: SO WHAT?

1. Wilson, E. O., *Sociobiology, The New Synthesis*, Cambridge, Mass., Belknap Press, 1975 – hereafter referred to as *Sociobiology*.

2. Wilson, E. O., *On Human Nature*, Cambridge, Mass., Harvard University Press, 1978 – hereafter referred to as *OHN*.

3. Marshall Sahlins uses this term in his book, *The Use and Abuse of Biology: An Anthropological Critique of Sociobiology*, London, Tavistock, 1977.

4. Trade advertisement in *Publisher's Weekly*, June 1978.

5. Publishing being what it is, I have had to prepare this chapter before the publication of *On Human Nature*, and am relying on an advance copy kindly lent to me by Professor John Maynard Smith; and a proof copy sent to us by Professor Edward Wilson himself.

6. *Sociobiology*, p.595; *OHN*, p.222.

7. Genes are the basic hereditary units. They contain all the information (in coded form) necessary to organise the development of the animal/plant. Thus 'a gene for thick fur', which would facilitate survival among animals living in cold regions, means that among the genes possessed by an animal is one that contains the information necessary to make the animal grow thick fur. This gene can then be passed on to the next generation at conception, so that the offspring are also better able to survive cold conditions. Some behaviour patterns are 'coded for' in the same way as anatomical features.

8. Dawkins, R., *The Selfish Gene*, Oxford University Press, 1976.

9. *Sociobiology*, p.550.

10. Your offspring carry only half your genes, because at fertilisation your genes join up with those of the individual with whom you have mated. Each egg and each sperm carry only half the original full 'set' of genes, when they are mature, precisely because of this joining up. If a mature sperm fertilised an egg, and both contained a full 'set' of genes, then the number would double in each generation.

11. Maynard Smith, J., *The Evolution of Sex*, Cambridge University Press, 1978.

12. *OHN*, p.125.

13. Dawkins, op. cit.

14. Quoted in Reynolds, V., *The Biology of Human Action*, Reading, Freeman, 1976.

15. *OHN*, p.20.

16. Clutton Brock, T. H. and Harvey, P. H. (eds), *Readings in Sociobiology*, Reading, Freeman, 1978; esp. Clutton Brock and Harvey, 'Primate Ecology and Social Organisation', pp.342ff.

17. Trivers, R. L., 'Parental Investment and Sexual Selection', in Clutton Brock and Harvey, op. cit., pp.52ff.

18. ibid., p.83.

19. Clutton Brock and Harvey, 'Evolutionary Rules and Primate Societies', in Clutton Brock and Harvey, op. cit., pp.293ff.

20. Teitelbaum, M. (ed.), *Sex Differences: Social and Biological Perspectives*, New York, Anchor Books, 1976; see esp. Lancaster, J., 'Sex Roles in Primate Societies', p.24.

21. *OHN*, p.31.

22. Ardrey, R., *The Hunting Hypothesis*, London, Collins, 1976; and see Reynolds, op. cit.

23. Tiger, L., *Men in Groups*, New York, Vintage Books, 1970.

24. Reiter, R. (ed.), *Toward an Anthropology of Women*, New York and London, Monthly Review Press, 1975; see esp. Slocum, S., 'Woman the Gatherer: Male Bias in Anthropology', pp.36ff.

25. *OHN*, p.34.

26. *OHN*, p.26.

27. Sahlins, M., *Stoneage Economics*, Chicago, Aldine, 1972.

28. E. P. Thompson, in *The Making of the English Working Class* (Harmondsworth, Penguin 1977), suggests that this was in fact the work ethic of pre-industrial society – that only with the rise of capitalism was work transformed into the pattern that is seen as 'natural' today, with clocking in, the 'working day', shift work, etc.

29. *OHN*, p.34.

30. Wilson, E. O., article in *New York Times Magazine*, 12.10.1975.

31. The change in diet was primarily to one with a higher meat content than that of other primates – which is not the same as saying that humans became 'meat-eaters'. This change itself has been questioned by Jolly, who has speculated that the first dietary shift in human evolution was to seed-eating, rather than to meat. If this is true, it will require some adjustment in theories about human evolution that rely on the 'hunting hypothesis'. Jolly, C., 'The Seed Eaters: A New Model of Hominid Differentiation Based on a Baboon Analogy', *Man*, 5 (1970), pp.5–26.

32. *Sociobiology*, p.563.

33. Slocum, op. cit., p.49.

34. A similar survey by Martin and Voorhies of 90 such societies also showed that, on average, meat forms only 30–40 per cent of the diet. In 58 per cent of the societies gathering was the primary subsistence activity, in comparison with 25 per cent where hunting predominated. Martin, M. K. and Voorhies, B., *The Female of the Species*, New York and London, Columbia University Press, 1975.

35. B. Hiatt, quoted in Brown, J., 'Anthropological Perspectives on Sex Roles in Subsistence', in Teitelbaum, op. cit., p.130.

36. Turnbull, C., *The Forest People*, London, Picador, 1976.

37. Goodale, J., *Tiwi Wives*, Seattle and London, University of Washington Press, 1971.

38. Friedl, E., *Women and Men: an Anthropologist's View*, Basic Anthropology Units. New York, Holt, Rinehart and Winston (1975).

39. Draper, P., 'IKung Women: Contrasts in Sexual Egalitarianism in Foraging and Sedentary Contexts', in Reiter, op. cit.

40. Beach, F. A., 'Human Sexuality and Evolution', in Coutinho, E. M. and Fuchs, F. (eds), *Physiology and General Reproduction*, New York, Plenum, 1974.

41. *OHN*, p.123.

42. *Sociobiology*, p.569.

43. Brown, op. cit.; see also Friedl, op. cit.

44. Friedl, op. cit.

45. Draper, op. cit.

46. Tiger, op. cit.

47. That females learnt to share with their mates first is also an assumption rather than a 'fact'. Sally Slocum speculates that it was more likely that females first extended the period of food-sharing with their children, with whom they were in intimate social contact over the many years of infancy; and that males first brought food home not to their spouses but to their mothers. This has considerable implications for theories of the origin of the nuclear family and the father role.

48. *OHN*, p.83.

49. Brown, op. cit.

50. *OHN*, pp.36ff.

51. *OHN*, p.37.

52. Rubin, G., 'The Traffic in Women: Notes on the "Political Economy of Sex"', in Reiter, op. cit., p.177.

53. Dawkins, op. cit.

54. Darwin, C., *The Origin of Species* (1859), Harmondsworth, Penguin, 1970.

55. Summary by Brown, op. cit., p.126; see also Fee, E., 'Science and the

Woman Problem', in Teitelbaum, op. cit., pp.126ff.

56. Tiger, L.; quoted in Martin and Voorhies, op. cit.

57. A word about the use of the term 'innateness'. Wilson uses the term in a much looser way than is usual in scientific writing. John Maynard Smith, in a review of *OHN* in *Nature*, writes: 'when [Wilson] speaks of a trait as innate, he does not mean that it will develop in a fixed way regardless of the environment. He means only that a trait will develop in rather a constant manner despite wide changes in the environment while accepting that a sufficiently drastic change in upbringing (or perhaps a sufficiently early one) might produce a profound change in outcome. By saying that characteristics are innate he means only that we acquire some characteristics very readily, and others only with great difficulty.' The near universality of certain sex differences would therefore identify them as being 'innate' in being particularly resistant to change.

58. *OHN*, pp.132ff.

59. ibid., p.127.

60. Children who, while genetically 'female', have been exposed to male hormones during their development in the womb, and reportedly show some 'masculinised' behaviour. Money, J. and Ehrhardt, A., *Man and Woman, Boy and Girl*, Baltimore, Johns Hopkins University Press, 1972.

61. Smith, review of *OHN* in *Nature*, op. cit.

62. Gould, S., 'Flaws in a Victorian Veil', *New Scientist*, Vol. 79, No. 1118, 31.8.78, p.632.

63. *OHN*, p.132.

64. Wilson, E. O., article in *New York Times Magazine*, 12.10.1975.

65. Tiger, L. and Shapher, J., *Women in the Kibbutz*, New York and London, Harcourt, 1975.

66. Blumberg, R. L., 'Kibbutz Women', in Iglitzin, L. B. and Ross, R. (eds), *Women in the World: A Comparative Study*, Santa Barbara, C. A., Clio Books, 1976.

67. Mead, M., *Sex and Temperament in Three Primitive Societies*, London, Routledge and Kegan Paul, 1977; see also Ashley Montagu (ed.), *Learning Human Non-Aggression*, New York, Oxford University Press, 1978.

68. Owen, L., 'The Myth of the Hairy Lady', *The Guardian*, 28.7.1977, p.11.

 * For those who would like to read further on the subject, I suggest *Signs:* Winter 1978. Volume on Women & Science contains some useful articles which follow the theme of this chapter. Caplan, Arthur, (ed.) *The Sociobiology Debate*, Harper & Row (1979): a collection of readings covering the debate.

CHAPTER 4
THE TYRANNICAL WOMB:
MENSTRUATION AND MENOPAUSE

1. Weideger, P., *Menstruation and Menopause*, New York, Arthur A. Knopf, 1977 (also published as *Female Cycles* in 1979 by A Women's Press, London).

2. Veith, I., *Hysteria: The History of a Disease*, Chicago, Phoenix Books/University of Chicago Press, 1970.

3. Bullough, V. and Voght, M., 'Women, Menstruation, and Nineteenth Century Medicine', *Bulletin of the History of Medicine*, 47 (1973), pp.66–82.

4. Ehrenreich, B., 'Gender and Objectivity in Medicine', *International Journal of Health Services*, 4 (4) (1974), pp.617–23.

5. Dr Archibald Church; cited by Barker-Benfield, B., 'Sexual Surgery in Late-Nineteenth Century America', *International Journal of Health Services*, 5 (2) (1975), pp.279–98.

6. ibid., p.290.

7. The development of this attitude is discussed in Ehrenreich, B., and English, D., *For Her Own Good: 150 Years of the Experts' Advice to Women*, New York, Anchor Press, 1978.

8. Barker-Benfield, B., *The Horrors of the Half-Known Life: Male Attitudes Toward Women and Sexuality in Nineteenth Century America*, New York, Harper and Row, 1976, pp.80–132.

9. The effects of the hormones in differentiating the two sexes, and in the menstrual cycle, are discussed briefly in the Appendix. We should also point out that Chinese doctors of the twelfth century AD knew something of hormone action, but this knowledge did not reach the West. Needham, J. and Gwei-Djen, L., 'Sex Hormones in the Middle Ages', *Endeavour*, 27 (1968), pp.130–2.

10. Ehrenreich, B. and English, D. *Complaints and Disorders: the Sexual Politics of Sickness*, Compendium, London, 1974, p.35.

11. Weideger, op. cit., discusses in detail the existence of the menstrual taboo.

12. Turnbull, C., *The Forest People*, London, Picador, 1976.

13. See for example, Shuttle, P. and Redgrove, P., *The Wise Wound: Menstruation and Everywoman*, London, Victor Gollancz, 1978. Also see The Matriarchy Study Group, *Menstrual Taboos* (pamphlet), London, Community Press, 1976.

14. Shuttle and Redgrove, op. cit., discuss at length the idea of menstruation as impurity; see in particular Part II.

15. Weideger, op. cit., p.102.

16. ibid., pp.95–103.

17. Bullough and Voght, op. cit., p.67.

18. Rich, A., *Of Woman Born: Motherhood as Experience and Institution*, London, Virago, 1977, p.106.

19. The extent and power of the taboo are brought home forcefully by considering – as Greer does in *The Female Eunuch* (London, Macgibbon and Kee, 1970) – that we will cheerfully suck a bleeding finger, but the idea of tasting menstrual blood fills most of us with abhorrence and loathing. Why? 'For us blood is part of living . . . [not a] sign of weakness'. (Rooney, F., 'Womanblood' in *The Lesbian Reader*, California, Amazon Press, 1975). Yet we cannot accept it.

20. Weideger, op. cit., Chapter 4.

21. McClintock, M., 'Menstrual Synchrony and Suppression', *Nature (London)*, 229 (1971), p.244.

22. There are a few exceptions. Weideger's book (op. cit.) and the book by Shuttle and Redgrove (op. cit.) are about menstruation and the beliefs surrounding it, for example.

23. We make no apologies for the heterosexuality and passivity of the term: it *is* a term frequently used in biological literature. No comment.

24. Shuttle and Redgrove, op. cit., refer to this possibility (although it is one that has appeared several times in the scientific literature – see the discussion in Chapter 8 below).

25. Short, R. V., 'The Evolution of Human Reproduction', *Proceedings of the Royal Society of London*, 195 (1976), pp.3–24.

26. e.g., London, N. B., Foxwell, M., Potts, D. M., Guild, A. L. and Short, R. V., 'Acceptability of an Oral Contraceptive that Reduces the Frequency of Menstruation: The Tri-cycle Pill Regimen', *British Medical Journal*, 2 (1977), pp.487–90.

27. Rich, op. cit., p.107.

28. Harding, M. E., *Woman's Mysteries*, New York, Bantam Books, 1973, pp.78–9.

29. The images of the terrible mother/good mother are discussed in Shuttle and Redgrove, op. cit.; see especially Chapter 5.

30. For a discussion of the role of the sun goddess, see Smart, N., *The Religious Experience of Mankind*, London, Fontana, 1971. She was called Ameratasu, and was important in ideas of purity and pollution: pp.155–6.

31. The word 'lunatic' derives from the Latin, *luna* meaning moon. It is interesting to note that the origin of another word indicating madness – hysteria – comes from the Greek for womb.

32. We have used the word 'changingness' deliberately. The more correct 'changeableness' implies an unpredictability, while 'changingness' conveys an image of changing in a more predictable way. We think women see themselves and their cycles as fairly predictable.

33. Hormones may have some effects on behaviour in humans, but, in general, the effects are slight. The male sex drive, for example, is to some extent affected by androgen levels. However, there is little evidence of direct *determination* of behaviour by hormones in the human species: that is, that in the absence of the hormone the behaviour does not appear, and in the presence of the hormone the behaviour appears. Yet popular books continue to be published which give the impression that, for instance, male aggression is directly determined by androgens. It is nothing of the sort, although it *may* be facilitated by androgens in a few individuals.

34. Money, J. and Ehrhardt, A. A., *Man and Woman, Boy and Girl*, Baltimore, Johns Hopkins University Press, 1972, p.223.

35. Shuttle and Redgrove, op. cit., p.89.

36. Janiger, O., Riffenburg, R., and Kersh, R., 'Cross-cultural study of premenstrual symptoms', *Psychosomatics*, 13 (1972), pp.226–35.

37. Dusser, D. B., and Gibbs, F. A., 'Variations in the Electro-encephalogram during the Menstrual Cycle', *Amer. J. Obstets. & Gynaecol.*, 44 (1942), pp.687–90.

38. Southam, A. L. and Gonzaga, F. P., 'Systemic Changes During the Menstrual Cycle', *Amer. J. Obstets. & Gynaecol.*, 91 (1965), pp.142 *et seq*.

39. Vierling, J. S. and Rock, J., 'Variations in Olfactory Sensitivity to Exaltolide during the Menstrual Cycle', *Journal of Applied Physiology*, 22 (1967), pp.311–15.

40. See for example, the studies by Moos and his colleagues: Moos, R. H. *et al.*, 'Fluctuations in Symptoms and Moods during the Menstrual Cycle', *Journal of Psychosomatic Research*, 13 (1969), pp.37–44; and Moos, R. H., 'Typology of Menstrual Cycle Symptoms', *Amer. J. Obstets & Gynaecol.*, 103 (1969), pp.391–402.

41. Parlee, M. B., 'The Premenstrual Syndrome', *Psychol. Bulletin*, 80 (1973), pp.454–65.

42. Parlee, M. B., 'Stereotypic Beliefs about Menstruation: a Methodological Note on the Moos Questionnaire and Some New Data', *Psychosomatic Medicine* 36 (1974), pp.229–40. Also see Parlee, M. B., 'Social Factors in the Psychology of Menstruation, Birth, and Menopause', *Primary Care*, 3 (3) (1976), pp.477–96.

43. Quotation given by Karen Paige in her article, 'Women Learn to Sing the Menstrual Blues', *Psychology Today*, September 1973.

44. For studies of academic performance and menstruation, see Bernstein, B. E., 'Effects of Menstruation on Academic Performance among College Women', *Archives of Sexual Behaviour*, 6 (4) (1977), pp.289–96; and Brooks, J., Ruble, D., and Clark, A., 'College Women's Attitudes and Expectations concerning Menstrual-related Changes', *Psychosomatic Medicine*, 39 (5) (1977), pp.288–98.

45. For example, see Broverman, D. M., Klaiber, E. L., Kobayashi, Y., and Vogel, W., 'Roles of Activation and Inhibition in Sex Differences in Cognitive Abilities', *Psychological Review*, 75 (1968), pp.23–50.

46. This attitude is typified by articles in popular magazines and newspapers. A good example is afforded by numerous articles in the mid-1970s that claimed that doctors had found that women in management positions tended to develop facial hair – i.e., they were becoming 'masculinised'.

47. This theory is developed by Dalton, K., *The Menstrual Cycle*, London, Penguin, 1969.

48. Janowsky, D., Berens, S. C., and Davis, J. M., 'Correlations between Mood, Weight, and Electrolytes during the Menstrual Cycle', *Psychosomatic Medicine*, 35 (1973), pp.143–54.

49. Horrobin, D. F., *Prolactin: Physiology and Clinical Significance*, Lancaster, MTP Press, 1973.

50. That is, a few women feel better, of those who felt sufficiently bad with premenstrual distress that they sought medical help. Most women who experience milder forms of distress do not seek medical help. Thus the proportion who do feel better after taking hormones is quite low. See for example, 'Recent Research on the Premenstrual Tension Syndrome', *Current Medical Research and Opinion*, 4, Suppl. 4 (1977), pp.9–40.

51. Sampson, G. A. and Jenner, F. A., 'Studies of Daily Recordings from the Moos Menstrual Distress Questionnaire', *British Journal of Psychiatry*, 130 (1977), pp.265–71.

52. The complex set of changes referred to as 'premenstrual tension' varies greatly from individual to individual, and often from cycle to cycle within the individual. Furthermore, since it is also culturally variable (see n.36 above), we find it absurd to attribute it simply to women's biology. If it were a direct consequence of our biology, we might expect it to be more constant in form.

53. A number of women claim to feel more energetic and creative at mid-cycle. See the discussion in Shuttle and Redgrove, op. cit., p.113.

54. Supposedly. Male constancy is probably also a myth. There has not been much research, but what there has been indicates that men too have cycles – of emotionality, for example, or of committing violent crimes. (The werewolf, you will remember, murdered whenever the moon was full.) See Hersey, R. B., 'Emotional Cycles in Man', *Journal of Mental Science*, 77 (1931), pp.151–69; and Lieber, A. and Sherin, C., 'The Case of the Full Moon', *Human Behaviour*, 1 (1972), p.29.

55. Shuttle and Redgrove, op. cit., pp.59–60.

56. Harding, op. cit., pp.93–4.

57. Bart, P., 'Depression in Middle-Aged Women', in Gornick, V. and

Moran, B. K. (eds), *Woman in Sexist Society*, New York, Basic Books, 1971.

58. Van Keep, P. A., and Humphrey, M., 'Psychosocial Aspects of the Climacteric', in Van Keep, P. A., Greenblatt, R. B., and Albeaux-Fernet, M. (eds), *Consensus on Menopause Research*, London, Medical and Technical Publishing, 1976.

59. Weideger, op. cit., p.210.

60. Brotherton, J., *Sex Hormone Pharmacology*, London, Academic Press, 1976.

61. These symptoms are precipitated very rapidly if the ovaries are removed with the womb in a hysterectomy – a 'surgical menopause'. This used to be standard procedure in hysterectomy and is still done in many cases, unless a woman requests otherwise. Obviously, ovaries should be removed if they are diseased, but it would be much better to leave them there if they are healthy; and women can then have a more gradual, natural menopause, which is less emotional and less of a physiological strain.

62. Weideger, op. cit., p.209.

63. This involves, for example, the growth of facial and body hair. It results from certain adrenal hormones which are masked by the high levels of oestrogen during the reproductive years.

64. Lebech, P. E., 'Effects and Side-effects of Estrogen Therapy', in van Keep *et al.*, op. cit.

65. Cooper, W., *No Change*, London, Hutchinson, 1975.

66. Cooper, W., 'The History of the Menopause', *Current Medical Research and Opinion*, 3, Suppl. 3 (1975), pp.3–18.

67. Thomson, J., and Oswald, I., 'Effect of Oestrogen on the Sleep, Mood and Anxiety of Menopausal Women', *British Medical Journal*, 2 (1977), pp.1317–19.

68. ibid. We use the term 'deficiency' in inverted commas because it is only defined as such by reference to levels produced in young, fertile women. Since a decline in oestrogens occurs in *all* women at the menopause, it seems an inappropriate word to use. The medical literature uses it most of the time. To give one example: 'The concept of normality of the menopause is being replaced by one of a chronic deficiency state lasting for up to 20 years.' From Studd, J., Chakravarti, S., and Oram, D., 'Practical Aspects of Hormone Replacement Therapy', *Current Medical Research and Opinion*, 3 Suppl. 3 (1975), pp.56–63.

69. By long periods we mean several years at least. This is very different from a woman who takes HRT for, say, a year to tide her over the worst menopausal problems. Unfortunately, many advocates of HRT encourage women to stay on it for years on end. The possible effects of this on health are completely unknown.

70. Gambrell, R. D., 'Estrogens, Progestagens, and Endometrial Cancer', in Van Keep *et al.*, op. cit.

71. Report of recommendations from the findings of the Royal College of General Practitioners: 'Oral Contraceptive Study on the Mortality Risks of Oral Contraception Users', *British Medical Journal*, 8.10.1977, p.947.

72. Weideger, op. cit., p.81.

73. Lebech, op. cit.

74. Birke, L., 'The Health of Half of Heaven', *China Now*, November/December 1978.

75. At least for as long as you go on taking it. More cautious doctors only prescribe HRT for short periods, while menopausal symptoms are particularly unpleasant. What concerns us here is the prescribing of HRT for many years. We have heard one advocate of HRT gleefully inform the audience that she knew of one old lady of ninety who was still on HRT, having been taking it for some time. We were told that she was still sprightly – thus implying that this had something to do with the drugs she was taking.

CHAPTER 5
FROM ZERO TO INFINITY:
SCIENTIFIC VIEWS OF LESBIANS

1. The word 'homosexuality' is a generic term derived from the Greek 'homo' meaning 'same'. 'Lesbian' derives from the isle of Lesbos, on which the famous lesbian poet, Sappho, once lived.

2. Tripp, C. A., *The Homosexual Matrix*, London, Quartet Books, 1977.

3. To illustrate this, I counted the number of new papers published on homosexuality, in which the sex of 'patient' was specified in the title, from *Index Medicus* listings for one full year (1977).

4. Wolff, C., *Love Between Women*, London, Duckworth, 1971.

5. Szasz, T., *The Manufacture of Madness*, London, Routledge & Kegan Paul, 1971.

6. Ehrenreich, B. and English, D., *Witches, Midwives and Nurses*, London, Compendium, 1971.

7. See Evans, A., *Witchcraft and the Gay Counterculture*, Boston, Fag Rag Books, 1978, for a discussion of what little is known about lesbians and the witch-hunts. He also discusses how traditional academic history has ignored the question of gay women and men being involved in the massacres.

8. Evans, op. cit., p.92.

9. Genital mutilation in nineteenth-century America is discussed, for example, in Ehrenreich, B. and English, D., *For Her Own Good: 150 Years of the Experts' Advice to Women*, New York, Anchor Press. 1978. It is also discussed by Barker-Benfield, G. J., *The Horrors of the*

Half-Known Life: Male Attitudes toward Women and Sexuality in Nineteenth-Century America, New York, Harper & Row, 1976. Genital mutilation in present-day African societies is also discussed in Hanmer, J. and Allen, P., 'Reproductive Engineering – the Final Solution?' (Chapter 10 below), and in Daly, M., *Gyn/Ecology: the Metaethics of Radical Feminism*, Boston, Beacon Press, 1978; see especially Chapter 5.

10. Freud, S., *Three Essays on the Theory of Sexuality* (1905), Harmondsworth, Pelican, 1973. Freud debated the question of an inherent psychological bisexuality in human beings. Like most other writers who mention homosexuality, however, he dealt mainly with the possible development of *male* homosexuals.

11. See, for example, Comfort, A., *The Anxiety-Makers: Some Curious Sexual Preoccupations of the Medical Profession*, London, Panther Books, 1968.

12. One can find many instances of letters to the press, both in the United States and in Europe, expressing horror at the thought of gay teachers or nurses. 'It would be kinder to the children if a teacher left the profession as soon as he or she thinks they are gay' is a common theme. Similar disgust was expressed over the idea that lesbians could want to have a child by artificial insemination – an article given wide publicity by the gutter press.

13. Kinsey, A. C., Pomeroy, W. B., Martin, C. E., and Gebhard, P. H., *Sexual Behavior in the Human Female*, Philadelphia, Saunders, 1953.

14. Hite, S., *The Hite Report*, London, Talmy Franklin, 1977.

15. Ford, C. S., and Beach, F. A., *Patterns of Sexual Behavior*, London, Eyre & Spottiswoode, 1952. The authors cite anthropological data available to them at the time of writing. Although many more papers have appeared since then, no one has seriously disputed the figure given by Ford and Beach as far as I know. What is changing more rapidly now are data on societies that condone lesbian behaviour among females, much of which was unknown until quite recently.

16. Shepherd, G., 'Lesbians in Mombasa', ms. of paper given at 'Women and Illness' conference, University of Sussex, Summer 1977.

17. Wikan, U., 'Man Becomes Woman: Transexualism in Oman as a Key to Gender Roles', *Man*, 2 (2) (1977), pp.304–19.

18. That is, androgens. These are sometimes called 'male' hormones because they predominate in males. Females, however, also produce them, but in much smaller quantities. The effects of hormones in differentiating the two sexes are described briefly in the Appendix.

19. Kreuz, L. E., Rose, R. M. and Jennings, J. R., 'Suppression of Plasma Testosterone Levels and Psychological Stress', *Archives of General Psychiatry*, 26 (1972), pp.479–82.

20. Some studies do report that a group of lesbians have higher androgen

levels than do heterosexual women, although many similar studies report no difference. The most comnonly cited evidence in favour of the view that lesbians have higher androgen levels comes from a paper by Loraine *et al.*, a paper that has come in for criticism (see n.30 below). The possibility that differences in hormonal levels result from some feature of behaviour, and not necessarily the other way round, does not often appear in the medical literature. It seems to be raised more often by biologists addressing questions of human behaviour.

21. It is theoretically quite possible (although there are no data to prove a point either way!) that relating sexually to men alters a woman's hormone output. It is certainly true that female hormone output can be affected in complex ways by the presence of males in some animal species. *If* this were true in women, then it may be that all women who are not relating sexually to men are comparable: this would include lesbians and celibate women.

22. Two examples of references to male-like behaviour as equivalent to lesbianism are: Dörner, G., *Hormones and Brain Differentiation*, Amsterdam, Elsevier, 1976 (the equation is made at various points in the book); and Herbert, J., 'Neurohormonal Integration of Sexual Behaviour in Female Primates', in Hutchison, J. B. (ed.), *Biological Determinants of Sexual Behaviour*, Chichester, John Wiley, 1978, p.487.

23. Dörner, op. cit., p.199.

24. Other means that have been suggested have included making tiny holes in the ventromedial nucleus of the hypothalamus (part of the brain dealing with several 'basic' functions such as eating, drinking, sex, etc.). Dörner (op. cit.) refers to the possibility of doing this on the basis of 'successes' (according to his criteria) in animal studies. He goes on to cite a similar result from human homosexual males in which lesions in the brain resulted in a reduction of the sex drive. They were therefore behaving less homosexually.

25. I am not saying here that there are no effects of the sex hormones on human behaviour. What I am saying is that there are few effects that show a *direct* effect of a hormone on behaviour; that is, that if you remove the source of the hormone the behaviour disappears, or if you increase the amount of the hormone the behaviour increases proportionate to the quantity of hormone. One of the few human behaviours that does seem to be fairly dependent on hormones (though not absolutely) is male sexual behaviour: castrated males rarely achieve an erection. Female sexual behaviour is also diminished, though not usually abolished, by removal of appropriate hormone sources (for women, this is the adrenal glands).

26. Kenyon, F. E., 'Female Homosexuality – a Review', in Loraine, J. A. (ed.), *Understanding Homosexuality*, London, Medical and Technical Publishing, 1974, p.86.

27. Swyer, G. I. M., 'Homosexuality: The Endocrine Aspects', *Practitioner*, 172 (1954), p.374.

28. The methods of obtaining data are crucial for scientific work. Criticism of work in science may rest on criticism of assumptions made by the scientist, the theories s/he puts forward, or the methods s/he uses. Usually, the attack is swift if a scientist's methods or theories are not 'up to standard'. It is curious, then, that those who write scientific papers about homosexuality rarely meet with much opposition, either from within science or from without.

29. Loraine, J. A., Adamopoulos, D. A., Kirkham, K. E., Ismail, A. A. A., and Dove, G. A., 'Patterns of Hormonal Secretion in Male and Female Homosexuals', *Nature, London*, 234 (1970), pp.552 *et seq.*

30. This study (Loraine *et al.*, op. cit.) involved taking urinary hormone samples, which is not thought to be a particularly good measure of hormone levels in blood, and comparing them to the *mean* of the control sample. The control sample were not drawn at random, but from workers at the Endocrinology Unit at Edinburgh. Thus it is at least as likely that any hormonal differences found reflected something about the workers at this unit as it does about the lesbian sample. Furthermore, the paper does not state where they got the lesbian sample from, but it does state that the women were in two couples. Since the menstrual cycles of women living together tend to synchronise (see Chapter 4 above), their hormone outputs may be similar or be affected by that of the partner. Control groups in studies such as this are often obtained from unknown sources too, or the groups are badly matched for social variables such as age, parity or socioeconomic class.

31. Eisenger, A. J., Huntsman, R. G., Lord, J., Merry, J., Polani, P., Tanner, J. M., Whitehouse, R. H., and Griffiths, P. D., 'Female Homosexuality', *Nature, London*, 238 (1972), p.106.

32. Siegelman, M., 'Adjustment of Homosexual and Heterosexual Women', *British Journal of Psychiatry*, 120 (1972), pp.477 *et seq.*

33. DeFries, Z., 'Pseudohomosexuality in Feminist Students', *American Journal of Psychiatry*, 133 (1976), pp.400–4.

34. I should point out that, while some contributors to the literature on homosexuality are fairly sympathetic to gays, they are often not sympathetic to women or to women's liberation!

35. West, D. J., 'Aspects of Homosexuality', in *Biosocial Aspects of Sex: Sixth Annual Symposium of the Eugenics Society, Journal of Biosocial Science* Suppl. 2, 1970.

36. West, D. J., *Homosexuality*, London, Duckworths, 1955.

37. Storr, A., *Sexual Deviation*, Harmondsworth, Penguin, 1964.

38. ibid.

39. West, op. cit.

40. Socarides, C. W., *et al.*, 'Homosexuality in the Male', *International Journal Psychiatry*, 11 (1973), pp.461–79.

41. I should point out that 'success' in such enterprises is a strange idea. I wonder how many of such patients are really frightened into 'accepting' a heterosexual life-style, while feeling that they would really prefer a homosexual life? Is this a success?

42. Chesler, P., *Women and Madness*, New York, Avon, 1972.

43. For example, Socarides, op. cit., and Bieber, I., 'Reply to Davison's [op. cit.] Article', *Journal of Consulting and Clinical Psychology*, 11 (1976), pp.163–6.

44. Davison, G. C., 'Homosexuality – the Ethical Challenge', *Journal of Consulting and Clinical Psychology*, 44 (1976), pp. 157–62.

45. ibid.

46. Socarides, op. cit.

47. Davison, op. cit.

48. The Sexual Offences Act (1967) decriminalised homosexual acts between consenting males over twenty-one in private. No other homosexual act between males was allowed under this Act, and so many arrests have been made of gay men since that time. Lesbians have always been exempt from legislation, although an attempt was made in 1921 to introduce an Act of Parliament. The reasons given for not passing the Act were based on politicians' fears that, by making it the subject of an Act, 'innocent' women would come to hear of lesbianism, and would be corrupted. See Weeks, J., *Coming Out: Homosexual Politics in Britain*, London, Quartet Books, 1977, pp.106–7.

49. Dörner, op. cit.; and also Dörner, G., 'Sex-hormone Dependent Differentiation in the Hypothalamus and Sexuality', in Lissak, K. (ed.), *Hormones and Brain Function*, London, Plenum Press, 1973.

50. Dörner, op. cit., (1976).

51. As an example he states that male gays have a 'female' brain response, in that their brains respond to an oestrogen injection by producing a pulse of LH (luteinising hormone) similar to that produced by female brains. At no point does he tell the reader how he defines a 'pulse': his criterion of a pulse in gay men appears to be much much smaller than that normally produced by females. Furthermore, if we take evidence from other primates, it may not be true that heterosexual males *never* respond to oestrogen by producing an LH pulse. For example, both male and female marmosets produce a pulse, and so do rhesus monkeys (see Hodges, J. K., and Hearn, J. P., 'A Positive Feedback Effect of Oestradiol on LH Release in the Male Marmoset Monkey, *Callithrix jacchus*', *Journal of Reproduction and Fertility*, 52 (1978), pp.83–6; and Karsch, F. J., Dierschke, D. J. and Knobil, E., 'Sexual Differentiation of Pituitary Function: Apparent Differences between Primates and Rodents', *Science* 179 (1973), pp.484–6). If this occurs in

other primates, it seems unlikely that human males would be different: Dörner, however, bases much of his work on the white rat. His woolliness would be attacked in any other branch of science. That it is not may be due to two factors: (1) homosexuality is still a taboo subject; (2) all this is embedded in a strictly scientific book, most of which is about copulatory postures of the rat, so it is accessible to few critics anyway.

52. The suicide rate in Britain is approximately 1.8 per cent. Dörner works in West Germany (Berlin); I would not expect, however, that the suicide rate in that country is significantly different.

53. Feldman, M. P., 'Aversion Therapy for Sexual Deviation: a Critical Review', *Psychological Bulletin*, 65 (1966), pp.65 *et seq.*

54. ibid.

55. Blitch, J. W., and Haynes, S. N., 'Multiple Behavioural Techniques in a Case of Female Homosexuality', *Journal of Behaviour Therapy & Experimental Psychiatry* 3 (1972), pp.319–22.

56. Interestingly, there are fewer references to the use of electric shock treatment. However, I have spoken to many lesbians who have been given it in the last ten years, either through the head or through the body, including the groin. They were usually told that it was for their lesbianism.

57. Rule, J., *Lesbian Images*, London, Peter Davies, 1976.

58. Abbott, S., and Love, B., *Sappho Was a Right-On Woman*, New York, Stein and Day, 1972.

59. From a poem, 'For straight folks who don't mind gays but wish they weren't so blatant', on the LP, 'Lesbian Concentrate: A Lesbian-thology'. The poem, by Pat Parker, points out that 'blatant heterosexuals are all over the place'. Olivia Records, 1977, LF 915. Also in *Womanslaughter*, an anthology of poems by Pat Parker, Oakland, California, Diana Press, 1978.

60. Some men may try to resist this, but it is still part of what men are taught in some societies, including our own. Rape and prostitution are particular manifestations of this expectation. Women are sometimes punished by rape if they refuse to 'give in'. Aspects of patriarchal control of women's bodies are discussed in Daly, op. cit. (see n.9 above).

61. I am also very much aware that the idea of two women making love is used time and time again to turn men on sexually – indeed, a great deal of porn is based on this. It represents yet another aspect of the way in which women are debased through their representation in porn, both 'soft' and 'hard'.

62. *A final note.* I am aware that I have used the term 'homosexuality' generically throughout this chapter. I am also aware that, whatever factors shape a person's development as a 'homosexual', these factors may well be different for lesbians and for gay men.

CHAPTER 6
'SICKNESS IS A WOMAN'S BUSINESS?':
REFLECTIONS ON THE ATTRIBUTION OF ILLNESS

1. 'Survey of Mental Health Statistics', *Mind*, July 1976.

2. Horney, K., *Feminine Psychology*, New York, Norton, 1973, p.231.

3. Oakley, A., *Sex, Gender and Society*, London, Maurice Temple Smith, 1972, p.16.

4. 'Survey of Mental Health Statistics', op. cit.

5. An epidemiologist is a medical specialist dealing with the rates of incidence of various diseases in different parts of the world. It comes from the same word-root as *epidemic*.

6. McMahon, B., Pugh, T. F., and Ipsen, J., *Epidemiologic Methods*, London, Churchill, 1960, p.105.

7. 'Depression' means different things to different people. Á simple distinction can be made between 'neurotic depression' and 'psychotic depression'.

   A neurosis implies mental conflict. It is a reflection of inner tensions. *The character of reality is the same for the neurotic*. He [sic] is fairly close to normality and people are usually tolerant of his symptoms because of the similarity between himself and others. Neurosis is the result of conflict between the individual and society and of conflicts within the individual. . . . *The very nature of reality changes for the psychotic*. . . . The symptoms of [psychotic] depressive conditions vary from total despair and brooding silence to irrational elation, constant . . . activity and unexplained irritability. [Emphasis in original] From Ennals, D., *Out of Mind*, London, Arrow Books, 1973.

8. Ngubane, H., *Body and Mind in Zulu Medicine*, London, Academic Press, 1977, p.155.

9. Last, M., in London, J. (ed.), *Social Anthropology and Medicine*, London, Academic Press, 1976.

10. Dalton, K., *The Menstrual Cycle*, Harmondsworth, Penguin, 1969.

11. Oakley, op. cit., pp. 44–5.

12. Janiger, O., Riffenburgh, R. and Kersh, R., 'Cross-Cultural Study of Premenstrual Symptoms', *Psychosomatics*, 13 (1972), pp.226–35.

13. Brown, G., Bhrolchain, Ni and Harris, T., 'Social Class and Psychiatric Disturbance among Women in an Urban Population', in *Sociology*, 9 (1975), pp.225–54.

14. ibid, pp.226–7.

15. Good, B., 'Semantics of Illness in Iran', *Culture, Medicine and Psychiatry*, 1 (1977), pp.25–58.

16. Nichter, M., 'Expression of Anxiety among Havik Brahman Women', in Standing, H. (ed.), *Women and Illness*, London, Academic Press, 1980.

17. Our category of psychosomatic illnesses is really a mystical one. It means illnesses of mind (psyche)-body (soma); that is, very often those complaints for which no obvious organic cause can be found or which are thought to be caused by states of mind. Migraine and backache are common examples. Although very real to the people who suffer them, such complaints may be dismissed as trivial or imaginary and contribute to stereotypes about, especially, the hypochondria of women.

18. Field, M., *Search for Security*, London, Faber, 1960.

19. ibid., p.149.

20. Bart, P. B., 'Depression in Middle-Aged Women', in Bardwick, J. (ed.), *Readings on the Psychology of Women*, New York, Harper and Row, 1972.

21. Wolf, M., *Women and the Family in Rural Taiwan*, Stanford University Press, 1972.

22. ibid., p.163.

23. ibid.

24. Bart, op. cit.

25. Last, op. cit.

26. Brooks, S., Ruble, D. and Clark, A., 'Menstrual-Related Expectations and Attitudes', *Psychosomatic Medicine*, 39(5), (1977), pp.288–98.

27. See especially *Menstrual Taboos*, Matriarchy Study Group Pamphlet; Shuttle, P., and Redgrove, P., *The Wise Wound: Menstruation and Everywoman*, London, Gollancz, 1978.

28. For those who would like to read further on this subject, I suggest Chesler, P., *Women and Madness*, London, Allen Lane, 1974; and Nathanson, C. A., 'Illness and the Feminine Role', *Social Science and Medicine*, 9 (1975), pp.57–62.

CHAPTER 7
THE OBSESSIVE ORGASM:
SCIENCE, SEX AND FEMALE SEXUALITY

1. Bodemer, Charles W., 'Concepts of Reproduction and its Regulation in the History of Western Civilisation', in *Contraception*, 13 (1976), pp.427–46.

2. Aberle, S. D. and Corner, G. W., *Twenty-five Years of Sex Research: History of the National Research Council Committee for Research in Problems of Sex*, London, Saunders, 1953.

3. Wandor, M. (ed.), *The Body Politic: Women's Liberation in Britain*

1969–72, London, stage 1 – see reference for Masters and Johnson, 'Human Sexual Responses', Boston, Little, Brown, 1966, quoted therein; also Kinsey, A. C. et al. *Sexual Behaviour in the Human Male* and *Sexual Behaviour in the Human Female*, Philadelphia, Saunders, 1953.

4. Bermant, G. and Davidson, J.: *Biological Basis of Sexual Behavior*, New York, Harper and Row, 1974, p.63.

5. ibid., p.67.

6. Ford, C. S. and Beach, F. A., *Patterns of Sexual Behavior*, London, Methuen, 1952; see also Pfaff *et al.*, 'Neurophysiological Analyses of Mating Behavior as Hormone-Sensitive Reflexes', *Progress in Physiological Psychology*, 5, (1973), pp.253–98.

7. Beach, F. A., 'Human Sexuality and Evolution', in Coutinho, E. M. and Fuchs, F. (eds), *Physiology and Genetics of Reproduction*, New York, Plenum Publishing, 1974.

8. Ford and Beach, op. cit., p.228; see also Doty, R. L., 'A Cry for the Liberation of the Female Rodent: Courtship and Copulation in Rodentia', *Psychological Bulletin*, 81 (1974), pp.159–72.

9. Beach, 1974, op. cit.; Ford and Beach, op. cit., p.171.

10. Beach, op. cit.

11. ibid.

12. Bermant and Davidson, op. cit., p.208.

13. ibid., p.210 (my emphasis).

14. ibid., pp.206–13

15. Oakley, A., *Sex, Gender and Society*, London, Maurice Temple Smith, 1975.

16. Hite, S., *The Hite Report*, London, Talmy Franklin, 1977.

17. ibid., pp.136ff.; Fisher, S., *Understanding the Female Orgasm*, Harmondsworth, Penguin, 1973, p.91. Fisher's study is conducted exclusively around intercourse and on married women. He does, however, find some correlation between women who prefer vaginal stimulation and high scores in 'depersonalisation' psychology tests. 'Depersonalisation' reflects a perception of your body as lacking vitality or arousal or "alive" quality' (p.97). These women are often less ecstatic or intense about emotions and body feelings though they often show higher levels of anxiety.

  Not only do these results totally contradict Freudian theories of female sexual maturity, but they may reflect the extreme effects of society's stereotypes. In other words, docility in women (or the desire to live 'happily ever after') and anxiety may be related to (false) expectations that men will bring all the satisfaction (both sexual and emotional) into our lives.

18. Hite, op. cit.

19. I use this to describe the psychologists, too numerous to name, who expect that the study of rats in particular can tell us something about human behaviour.

20. Shuttle, P. and Redgrove, P., *The Wise Wound: Menstruation and Everywoman*, London, Victor Gollancz, 1978, pp.88–9.

21. ibid., p.294ff.

22. ibid.

23. Sanders, L. and Wallis, J. (eds), *Conditions of Illusion: Papers from the Women's Movement*, Leeds, Feminist Books, 1974, p.89. See reference for Dr Sherfey, M. J., *The Nature and Evolution of Female Sexuality*, New York, Vintage Books, 1973. Quoted therein.

24. Oakley, op. cit., pp.109–11.

25. Ford and Beach, op. cit., pp.162–73.

26. ibid.

27. ibid.

28. Shuttle and Redgrove, op. cit., p.285.

29. Ford and Beach, op. cit., pp.162–73.

30. 'A woman will masturbate if she is sexually excited and there is no man to satisfy her. A couple may be having intercourse in the same house, or near enough for her to see them, and she may thus become aroused. She then sits down and bends her right leg so that her heel presses against her genitalia. Even young girls of about six years may do this quite casually as they sit on the ground. The women and men talk about it freely, and there is no shame attached to it.' Oakley, op. cit., p.104.

31. Ford and Beach, op. cit., pp.162–73.

32. Oakley, op. cit., p.104.

33. ibid., pp.162–73.

34. Beach, op. cit.

35. Bermant and Davidson, op. cit., p.67.

36. There is evidence that higher primates are less dependent on oestrus for sex than 'lower' mammals. Baboons exhibit swelling and reddening between menstruation and ovulation; 'old world' monkeys and apes in the less demanding or artificial environments of zoos, etc., are receptive longer than the period of ovulation (Hite, op. cit., pp.139–40).

37. Hite, op. cit., pp.139–40.

38. Beach, op. cit.

39. Wilson, E. O., *On Human Nature*, Cambridge, Mass., Harvard University Press, 1978, pp. 137, 140.

40. In Jewish tribes, for example, homosexuality and cunnilingus were an integral part of people's physical relations. When these tribes fought to separate and build their society as the 'Chosen People of God' and

confined their worship to one male god, these forms of sexuality were banned by religious code (Hite, op. cit., pp.142-3).

41. Hite, op. cit., pp.142-3.

42. Barker-Benfield, G. J., *The Horrors of the Half-Known Life: Male Attitudes Toward Women and Sexuality in Nineteenth-Century America*, New York, Harper and Row, 1976, pp.276-7.

43. ibid.

44. This role has continued into recent times. It was popular belief that masturbation led to insanity and mothers were obliged to go to enormous lengths to 'save their children'. 'Untiring zeal on the part of the mother or nurse is the only cure; it may be necessary to put the legs in splints before putting the child to bed' (Wandor, op. cit., p.200).

45. Comfort, A., *The Anxiety Makers*, London, Panther Books, 1968, p.46.

46. Barker-Benfield, op. cit., pp.277-9, 281-2.

47. ibid., pp.120-5.

48. ibid.

49. ibid., pp.128-9.

50. ibid.

51. ibid., pp.126-8.

52. ibid.

53. Rowbotham, S., *Woman's Consciousness, Man's World*, Harmondsworth, Penguin, 1974, p.283.

54. ibid., p.36.

55. Oakley, op. cit., p.99.

56. Fisher, op. cit., p.213.

57. Barker-Benfield, op. cit., p.277.

58. Hite, op. cit., p.138 (my emphasis).

59. Singer, S., *Androgyny: Towards a New Theory of Sexuality*, London, Routledge & Kegan Paul, 1977, p.272.

CHAPTER 8
TECHNOLOGY IN THE LYING-IN ROOM

1. Lisitzin, Y., *Health Protection in the USSR*, Moscow, Progress Publishers, 1972.

2 Hodgkinson, R., *Science and Public Health*, Open University Books, *Science and the Rise of Technology since 1800*, Block V, Unit, 10, 1973;

3. Woodham-Smith, C., *Florence Nightingale*, London, Fontana, 1964.

4. Graham, H., *Eternal Eve: The Mysteries of Birth and the Customs that Surround It*, London, Hutchinson, 1960.

5. Rich, A., *Of Woman Born: Motherhood as Experience and Institution*, London, Virago, 1977.

6. Hodgkinson, op. cit.

7. Kitzinger, S., *The Good Birth Guide*, London, Fontana, 1979.

8. *Report of the Peel Committee* (Standing Maternity and Midwifery Advisory Committee), London, HMSO, 1970.

9. It is strange to realise that only in the last 100 years or so has the medical *profession* had any appreciable positive effect on health. Before that, untested theories of physiology, and ignorance of the processes of infection, etc., meant that professional doctors often did more harm than good. See Donnison, J., *Midwives and Medical Men*, London, Heinemann, 1977.

10. *British Births 1970*, a survey directed by R. Chamberlain *et al.*, London, Heinemann, 1975.

11. *Annual Statistics*, Department of Health and Social Security, London.

12. Chalmers, I. and Richards, M., 'Intervention and Causal Interference in Obstetric Practice', in Chard, T., and Richards, M. (eds), *Benefits and Hazards of the New Obstetrics*, Clinics in Developmental Medicine 64 (1977).

13. ibid.

14. Tunstall, S., *Experiencing Childbirth: A Survey of 40 Islington Women*, a report of a survey carried out for Islington Community Health Council (unpublished ms.) 1978.

15. Rich, op. cit., p.159.

16. Kitzinger, S. and Davis, J. A. (eds), *The Place of Birth*, Oxford University Press, 1978.

17. Hippocrates and Galen of Pergamon were physicians of Classical times, who have had considerable influence on medicine throughout the ages.

18. Rich, op. cit., p.133.

19. Oakley, A., 'The Trap of Medicalised Motherhood', *New Society*, 34 (689) (1975), p.639–41.

20. Rich, op. cit.

21. ibid., p.185.

22. Graham, H. and McKee, L., 'Ideologies of Motherhood and Medicine on Radio and Television', paper presented to the British Sociological Association conference, University of Sussex, 1978.

23. ibid.

24. Breen, D., 'The Mother and the Hospital', in Lipshitz, S. (ed.), *Tearing the Veil: Essays on Femininity*, London, Routledge & Kegan Paul, 1978.

25. Turnbull, A. C., Introduction to Chard and Richards, op. cit.

26. Chard, T., and Richards, M., 'Lessons for the Future', in Chard and Richards, op. cit.

27. Radical Statistics Health Group, *In Defence of the NHS*, produced for the British Society for Social Responsibility in Science, London.

28. McManus, T. J. and Calder, A. A., 'Upright Posture and the Efficiency of Labour', *Lancet*, 1 (3055) (1978), pp.72–6; Dunn, P. M., letter to the *Lancet* challenging McManus and Calder, op. cit., *Lancet*, 1 (3062) (1978), p.496.

29. Rich, op. cit.

30. Tunstall, op. cit.

31. Brook, D., *Nature-birth: Preparing for Natural Birth in an Age of Technology*, Harmondsworth, Penguin, 1976.

32. M. P. M. Richards, 'A Place of Safety? An Examination of the Risks of Hospital Delivery', in Kitzinger and Davis, op. cit.

33. Rindfuss, R. S. and Ladinsky, J. L., 'Patterns of Births: Implications for the Incidence of Elective Induction', *Medical Care*, 14, (1976), pp.685 *et seq.*

34. It is precisely in terms of this stress on the welfare of the yet-to-be-born that opponents of the use of drugs in labour have argued: most analgesic techniques and induction agents used in birth have notable effects on the foetus. (See *British Births* 1970; Graham and McKee, op. cit.). For example, pethidine, an analgesic often used in delivery, can prolong the delay before the new-born baby takes its first breath. The use of pethidine has increased despite a much wider use of epidurals. Consequently, use of such drugs or techniques should be limited to those occasions when absolutely necessary on medical grounds. See Barden, T. P., 'Induction of Labour with Prostaglandins', in Ranwell, P. W. (ed.), *The Prostaglandins*, London, Plenum Press, 1977; *Spare Rib*, 73, August 1978 (London), pp.6ff.

35. Richards, op. cit.

36. McNay, M. B., McIlwaine, G. M., Howie, P. W. and MacNaughton, M. C., 'Perinatal Deaths: Analysis by Clinical Cause to Assess Value of Induction of Labour', *British Medical Journal*, 1 (1977), pp.347–50.

37. Yudkin, P., 'Problems in Assessing Effects of Induction of Labour on Perinatal Mortality', *Brit. J. Obstets. and Gynaecol.*, 33, (1976), pp.603–7; Turnbull, op. cit.

38. But in some regions of Britain perinatal mortality is still depressingly high – notably areas with a high working-class population. The rate is lower in Holland, for instance, and has been for some years – where home confinement for most women is encouraged.

39. Tunstall, op. cit.

40. Donnison, op. cit.

41. Riley, E. M. D., 'What Do Women Want? – The Question of Choice in the Conduct of Labour', in Chard and Richards, op. cit.

42. Tunstall, op. cit.

43. *Woman* magazine supplement: *How to Enjoy Having a Baby* (London), 22.4.1978.

44. ibid.

45. Brook, op. cit., Rich, op. cit.

46. Breen, op. cit.

47. Rich, op. cit., p.129.

48. This is not likely to be fully effective within the present health service structure. At present private medicine, contrary to the claims of the BMA, saps the facilities available on the National Health Service. Women can receive adequate care only within an adequate health service: and we will not have that until private medicine is abolished and medical practice fully socialised.

49. *Spare Rib* no. 73, op. cit.

50. ibid.

51. There are now a number of organisations supporting the moves against hospital confinement and the sometimes inhumane treatment women receive during labour. These include: Association of Radical Midwives, c/o Jen Flintham, 12a John St, Cambridge; Association for the Improvement of Maternity Services, c/o Christine Beels, 19 Broomfield Crescent, Leeds 6; Society to Support Home Confinements, c/o Margaret White, 17 Laburnum Ave, Durham; National Childbirth Trust, 9 Queensborough Terrace, London W2.

   Most of these do not attempt to challenge directly the *structure* of the health service, and its inequalities; organisations whose policies include a fully socialised health service include the Socialist Medical Association, 9 Poland St, London W1V 3DG, and Marxists in Medicine, 16 King St, London WC2, which produces the magazine *Medicine in Society*.

52. Kitzinger, S., 'A Note to the Reader', in *Education and Counselling for Childbirth*, London, Bailliere Tindall, 1977.

CHAPTER 9
CONTRACEPTION:
THE GROWTH OF A TECHNOLOGY

1. Ryder, N. B. and Westoff, C. F., *Science*, 153 (1966), p.1199.

2. In which Charles Bradlaugh and Annie Besant challenged a court decision by reprinting and selling copies of Knowlton's *Fruits of*

Philosophy, a study of the population question which included some advice for avoiding conception.

3. Peel, J. 'The Manufacture and Retailing of Contraceptives in England', *Population Studies* 17 (1963), p.113.

4. Vulcanisation (heating with sulphur to increase elasticity and strength) was invented in 1844, simultaneously in the United States and UK by Goodyear and Hancock, but it was not applied to contraceptives until the 1880s.

5. Peel, op. cit.

6. Peel, J. and Potts, M., *Textbook of Contraceptive Practice*, Cambridge University Press, 1969.

7. Ferguson, S., and Fitzgerald, H., 'Studies in the Social Services', in Hancock, H. (ed.), *History of the Second World War*, London, HMSO, 1954.

8. Parkes, A. S., *Sex, Science and Society*, London, Oriel Press, 1966.

9. Vaughan, P., quoted in 'The Story of the Pill', transcript of a BBC Radio 3 programme, 1971, produced by Robin Brightwell.

10. Marrian, G. F., also quoted in 'The Story of the Pill'; Parkes, A. S., 'Obituary of F. H. A. Marshall', *Endocrinology*, 6 (1948); Zuckerman, S., 'Oliver Bird Lecture', 1966, referred to in his *Beyond the Ivory Tower*, London, Weidenfeld & Nicolson, 1970; Corner, G. W., *The Hormones in Human Reproduction*, Princeton University Press, 1942.

11. Peel, J., 'Contraception and the Medical Profession', *Population Studies*, 18 (1964), p.133.

12. Routh, C. H. F., *Medical Press and Circular*, October 1878; quoted by Peel, op. cit.; also Smith-Rosenberg, C., and Rosenberg, C., 'The Female Animal: Medical and Biological Views of Woman and her Role in Nineteenth Century America', *Journal of American History*, (1973), p.332.

13. *The Practitioner*, July 1923, special issue on contraception.

14. Peel, 'Contraception', op. cit.

15. Shryock, R. H.: *The Development of Modern Medicine: An Interpretation of the Social and Scientific Factors Involved*, London, Gollancz, 1948.

16. *The Practitioner*, July 1923.

17. Mears, E., 'The Medical Student and Sex Education', paper to the International Planned Parenthood Conference, The Hague, June 1961.

18. Peel and Potts, op. cit.; Banks, J. A., *Prosperity and Parenthood*, London, Routledge & Kegan Paul, 1954.

19. Kennedy, D. M., *Birth Control in America*, New Haven, Yale University Press, 1970.

20. Sanger, M., and Stone, H. (eds), *The Practice of Contraception*,

Baltimore, Williams and Wilkins, 1931; and Sanger, M., *International Neo-Malthusian and Birth Control Conference*, vols I–IV, New York, American Birth Control League, 1926. Both read like catalogues of the inadequacies of existing methods.

21. Mears, op. cit.

22. Bernal, J. D., *Science and Industry in the Nineteenth Century*, London, Routledge, & Kegan Paul, 1953, pp.88–9.

23. Peel, 'Manufacture', op. cit.

24. Glass, D. V., *Population Policies and Movements in Europe*, Oxford University Press, 1940; Eldridge, H. T., 'Population Policy', in *International Encyclopaedia of the Social Sciences*, London, Macmillan, 1968.

25. Ferguson and Fitzgerald, op. cit.

26. Winter, I., *Journal of the American Medical Association*, 212 (1970), p.1067.

27. Peel and Potts, op. cit.

28. Ministry of Health (UK) *Memorandum*, 153/MCW, July 1930.

29. Peel, 'Contraception', op. cit.

30. The Royal Commission on Population (UK), *Report*, Cmnd 7695, London, HMSO, 1949.

31. Draper, E., *Birth Control in the Modern World*, Harmondsworth, Penguin, 1965.

32. Although in fact Memo 153 was incorporated into the 1948 National Health Service Act without the restriction 'married women', many clinics refused to see women who were not married, or about to be married, even in the late 1960s.

33. Family Planning Association, *Annual Report*, 1976.

34. Eldridge, op. cit.

35. ibid.

36. Symonds, R. and Carder, M., *The United Nations and the Population Question*, Brighton, Sussex University Press, 1973.

37. Petersen, W., *The Politics of Population*, London, Gollancz, 1964.

38. Pincus, G., *The Control of Fertility*, London, Academic Press, 1965.

39. Nelson, W. O., *Endocrinology*, 59 (1965), p.140.

40. *Science Newsletter*, 27.10.1951.

41. Symonds and Carder, op. cit.

42. Draper, op. cit.

43. *Congress and the Nation, A Review of Government and Politics*, vol. II, 1965–68, Congressional Quarterly Service, Washington, US Government Printing Office, 1969.

44. Freedman, R., 'Fertility', in *International Encyclopaedia of the Social*

Sciences, London, Macmillan, 1968; Wrigley, E. A., *Population and History*, London, Weidenfeld & Nicolson, 1969; Banks, op. cit.

45. Himes, N. E., *Medical History of Contraception*, Baltimore, 1936.

46. Peel, 'Manufacture', op. cit.; Banks, op. cit.

47. Freedman, op. cit.

48. ibid.

49. Banks, op. cit.

50. Freedman, op. cit.; Himes, op. cit.

51. Whelpton, P. K., Campbell, A. A. and Patterson, J. E., *Fertility and Family Planning in the United States*, Princeton University Press, 1965.

52. Chafe, W. H., *The American Woman, Her Changing Social, Economic and Political Roles 1920–1970*, New York, Oxford University Press, 1972.

53. Ferriss, A. L., *Indicators of Trends in the Status of American Women*, Russel Sage Foundation, 1971.

54. Banks, op. cit.

55. Chafe, op. cit.

56. Komarovsky, M., *Blue Collar Marriage*, New York, 1962.

57. ibid.

58. *Social Indicators, 1973*, Executive Office of the President/Office of Management of the Budget, Washington, US Government Printing Office, 1973, pp.189 *et seq.*; p.107.

59. United Nations, *Seminar on the Status of Women and Family Planning*, Istanbul, July 1972; report published New York, 1972 (Section III).

60. Whelpton *et al.*, op. cit.

61. Ryder and Westoff, op. cit.

62. Malthus, T. R., *Essay on the Principles of Population* (1798), Harmondsworth, Penguin, 1970.

63. His now famous proposition, first formulated in 1798, was that food and other resources increased only arithmetically, while population increased geometrically: in other words, that food supply increases only by steady increments, but population tends to double, double again, and so on.

64. *The Malthusian*, January 1906.

65. ibid., July 1906.

66. ibid., February 1909.

67. Dowse, R. E. and Peel, J., 'The Politics of Birth Control', *Political Studies*, 13, (1965), p.179.

68. There was no 'birth control movement' as such, but the term is used in this chapter to describe collectively organisations like the Malthusian

League and the Family Planning Association (established in 1939) in Britain, the Birth Control League of America (1921), the International Planned Parenthood Federation (1948), and individuals such as Margaret Sanger and Marie Stopes. From the 1920s onwards, many of the most well-known campaigners, as well as large numbers of the supporters of the birth control movement, were women, but it was not exclusively a women's movement.

69. Micklewright, F. H. A., 'The Rise and Decline of English Neo-Malthusianism', *Population Studies*, 15 (1961), p.32; also Fryer, P., *The Birth Controllers*, London, Secker & Warburg, 1965.

70. *Birth Control Hearings on HR 5978*, 73rd Congress, 2nd session, 1934, p.150, US Congress, House Committee on the Judiciary.

71. Hansen, J., *The Population Explosion*, Merit Books, 1970.

72. *Science for the People* (USA) 5, (2) pp.4–8: survey of the United States' use of population control in Latin America.

73. ibid.

74. ibid.

75. Drayton, H. A., *New Scientist*, 1.10.1970,

76. Kennedy, op. cit.

77. Peel, 'Contraception', op. cit.

78. Freeman, C., 'Malthus with a Computer', in Cole, H. S. D. *et al.* (eds), *Thinking About The Future*, Brighton, Sussex University Press, 1973.

79. Drayton, op. cit.

80. See n.68 above.

81. Peel, 'Contraception', op. cit.

82. Kennedy, op. cit.

83. Peel, 'Contraception', op. cit.

84. Kennedy, op. cit.

85. Weisner, B. P., 'Experiments of Controlled Fertility by Means of Hormonic Interference', in Blacker, C. P. (ed.), *International Medical Groups for the Investigation of Birth Control 1929 Report*; also Taylor, H., 'Report on Hormonic Control of Fertility', in Sanger and Stone, op. cit.

86. With the following difference: in the birth control movement were a number of supporters of the eugenics movement as well as those who supported a 'woman's right to choose' (as has been mentioned), whereas today's pro-abortion lobby is not associated with the now-discredited eugenics movement.

87. Kennedy, op. cit.

88. Peel, 'Manufacture', op. cit.

89. Lewis-Faning, H., *Report on an Enquiry into Family Limitation and its Influence on Human Fertility in The Past Fifty Years*, Papers of the

Royal Commission on Population, Vol. I, Cmnd 7695, London, HMSO, 1949.

90. Whelpton, P. K., and Kiser, C. V., (eds), *Social and Psychological Factors Affecting Fertility*, vol. 2, Milbank Memorial Fund, 1950, p.212.

91. Freedman, R., Whelpton, P. K., and Campbell, A. A., *Family Planning, Sterility and Population Growth*, New York, McGraw Hill, 1959.

92. Komarovsky, op. cit.

93. See Freedman, op. cit.

94. Kennedy, op. cit.

95. 'The Accident of Birth', *Fortune*, February 1938, p.83.

96. Voge, C. I. B., 'The Present Status of the Contraceptive Trade', *Manufacturing Chemist*, (1933), p.289.

97. Winter, op. cit.

98. British Drug Houses, in fact, had marketed the spermicide Volpar, which had been developed as a result of work by the Birth Control Investigation Committee, but it was not sold in the United States where Winter was writing.

99. Quoted by Vaughan, P., *The Pill on Trial*, London, Weidenfeld & Nicolson, 1970.

100. ibid.

101. Mintz, M., *The Pill*, Hodder-Fawcett, 1969.

102. Drill, V., *Oral Contraceptives*, McGraw Hill, 1966 (publisher's comments).

103. Mintz, op. cit.

104. ibid.

105. Oyediran, M. A., and Akinyanju, P. A., *Nigerian Medical Journal*, 6 (1976), p.464.

106. The process by which a clot forms inside a blood vessel and may then break off and travel around the body, eventually blocking a smaller blood vessel – for example, in the heart.

107. Vessey, M. P., and Inman, W. H. W., *British Medical Journal*, 2 (1968), p.193; also Vessey, M. P., and Doll, R., ibid., p.199.

108. Searle's first pill was called Enovid in the United States and Conovid in the UK. The UK launch was one to two years later than that in the United States.

109. These figures are approximate because the number of brands of pill and number of progestins in them continually changes, as new ones are introduced and others withdrawn.

110. Oyediran and Akinyanju, op. cit.

111. Rakusen, J., *Spare Rib*, 42, December 1975.

112. *The Sunday Times*, 18.1.1976.

113. Rakusen, J., *Spare Rib*, 47, June 1976; Harne, L., ibid., 69, April 1978; Shapiro, R., and Jones, D., *The Leveller*, 15, March 1978.

114. *New Scientist*, 29.4.1976; Lall, S., *Major Issues in Transfer of Technology to Developing Countries*, UNCTAD, 24.11.1975; The Haselmere Group: *Who Needs the Drug Companies?*, 1976.

115. Joint Working Group on Oral Contraceptives, *Report*, London, HMSO, September 1976.

116. *The Times*, 24.3.1976.

117. Practolol was given for heart complaints.

118. Dollery, C., and Rawlins, M., *British Medical Journal*, 8.1.1977.

CHAPTER 10
REPRODUCTIVE ENGINEERING:
THE FINAL SOLUTION?

1. Woolf, V., *Three Guineas* (1938), London, Harmondsworth, Penguin, 1977.

2. In this article we use the term 'reproductive engineering' in preference to 'genetic engineering', since we are concerned not with manipulation of the genes themselves, but with ways of interfering in the reproduction process from the formation of eggs and sperm to birth.

3. 'Genetic Engineering: Reprieve', *Journal of the American Medical Association*, 220 (10) (1972), 1355, 57, quoted in 'Genetic Engineering: Evolution of a Technological Issue', US House of Representatives, 92nd Congress, second session 1972; and see also 93rd Congress, second session 1974, Science Policy Research Division, Congressional Research Service, Library of Congress, US Government Printing Office.

   Geneticists now use the term 'genetic engineering' to mean certain techniques for studying DNA involving the transfer of genes from an organism to a bacterial host. This research is not without its social implications, but these have been well publicised and lie outside the scope of this paper. For a review, see Grobstein, C., 'The Recombinant DNA Debate', *Scientific American*, 237 (1977), pp.22–3.

4. We use the term 'sex predetermination' in preference to the liberal term, 'sex choice'. While it is obviously ridiculous to speak of sex choice for cattle, similar arguments can be made for women. How free is choice in a society where some people are more powerful than others and where social forces including the mass media create 'need'? 'Sex predetermination' conveys that the decision may be made by someone other than the mother.

5. Kass, L., 'Making Babies: The New Biology and the Old Morality', *Public Interest*, Winter 1972, pp.13–56.

6. Edwards, R., 'Fertilisation of Human Eggs in Vitro: Morals, Ethics and the Law', *Quarterly Review of Biology*, March 1974, pp.3–26.

7. Kass, op. cit., p.22.

8. Grossman, E., 'The Obsolescent Mother', *Atlantic*, 227 (1972), pp.39–50.

9. Firestone, S., *The Dialectic of Sex*, London, Jonathan Cape, 1971; and Chesler, P., *Women and Madness*, New York, Avon, 1972.

10. Hockley, W., 'Dysgenics – A Social Problem. Reality Evaded by the Illusion of Infinite Plasticity of Human Intelligence', *Phi Delta Kappa*, March 1972, pp.291–5.

11. Brudenell, M., McLaren, A., Short, R. and Symonds, M., *Artificial Insemination*, London, Royal College of Obstetricians and Gynaecologists, 1976.

12. Etzioni, E., 'Sex Control, Science and Society', *Science*, 13.9.1968, pp.1107–112; see also Williamson, N., *Sons or Daughters: A Cross Cultural Survey of Parental Preferences*, London and Beverley Hill, Sage, 1976.

13. Ju, K. S., Park, I. J., Jones, H. W. R. and Winn, K. S., 'Pre-natal Sex Determination by Observation of the X and Y Chromatin of Exfoliated Amniotic Fluid Cells', *Obstetrics and Gynaecology*, 47 (3) (1976), pp.287–90.

14. Rorvik, D. and Shettles, L., *Your Baby's Sex: Now You Can Choose*, London, Cassell, 1970; and *Choose Your Baby's Sex*, New York, Dodd, Mead, 1977.

15. Shastry, P. R., Hegde, U. C. and Rao, S. S., 'Use of Ficoll-Sodium Metrizoate Density Gradient to Separate Human X and Y Bearing Spermatozoa', *Nature*, 269 (1977), p.58.

16. US Sub-Committee, op. cit., p.176; 'Prospect for Genetic Intervention in Man', by Bernard D. Davis. The author is professor of bacterial physiology at Harvard Medical School.

17. Postgate, J., 'Bat's Chance in Hell', *New Scientst*, 5.4.1973, pp.12–16.

18. Etzioni, op. cit.

19. Edwards, Banister, and Steptoe, 'Early Stages of Fertilisation of Human Oocytes Matured in Vitro', *Nature*, 221 (1969), pp.632–35.

20. Firestone, op. cit.

21. Galana, L., 'Radical Reproduction: X Without Y', in Covina, G. and Galana L. (eds), *The Lesbian Reader*, Oakland, CA., Amazon Press, 1975, pp.122–37; Rose, H. and Hanmer, J., 'Women's Liberation, Reproduction and the Technological Fix', in Barker, D. and Allen, S., *Sexual Divisions and Society: Process and Change*, London, Tavistock, 1976, pp.199–223.

22. Rorvik, D., *In His Image*, London, Hamish Hamilton, 1978.

23. Gurdon, J. B., 'The Developmental Capacity of Nuclei Taken from

Intestinal Epithelium Cells of Feeding Tadpoles', *Journal of Embryology and Experimental Morphology*, 10 (1962), pp.622–40.

24. De Robertis, E. M. and Gurdon, J. B., 'Gene Activation in Somatic Nuclei after Injection into Amphibian Oocytes', *Proceedings of the National Academy of Science* (US), 74 (1977), pp.2470–4.

25. Modlinski, J. A., 'Transfer of Embryonic Nuclei to Fertilized Mouse Eggs and Development of Tetraploid Blastocysts', *Nature*, 273 (1978), pp.466–7.

26. Mittwoch, U., 'Virgin Birth', *New Scientist*, 78 (1107) (1978), 750–2; Kelly, J., 'Parthenogenesis', *Off Our Backs*, 7 (1) (1977). Both these articles review the field, and Kelly has an extensive bibliography.

27. We would like to thank Dilis O'Farrell for her help with this section.

28. Edwards, op. cit.

29. Gorney, R., *The Human Agenda*, New York, Bantam, 1973, p.221.

30. Hanmer, J., 'Violence and the Social Control of Women', in Littlejohn, C. *et al.* (eds), *Power and the State*, London, Croom Helm, 1978.

31. Ehrenreich, B., and English D., *Witches, Midwives and Nurses: A History of Women Healers*, Oyster Bay, NY, Glass Mountain Pamphlets, 1970.

32. Piers, M., *Infanticide*, New York, Norton, 1978; and Harris, M., *Cows Pigs Wars and Witches: The Riddles of Culture*, New York, Vintage, 1975.

33. Webster, P., 'The Politics of Rape in Primitive Society', *Michigan Discussions in Anthropology*, 1978.

34. Hoskin, F., 'Female Circumcision and Fertility in Africa', *Women and Health*, 1 (6) (1976).

35. Shandall, A., 'Circumcision and Infibulation of Females', *Sudan Medical Journal*, 5 (4) (1967).

36. Levy, H. S., *Chinese Footbinding: The History of a Chinese Exotic Custom*, New York, W. Rawls, 1966.

37. Morgan, R., 'Lesbianism and Feminism: Synonyms or Contradictions?' paper presented at a conference in Los Angeles, CA.; reprinted in *Going Too Far*, Vintage, 1978.

38. *Sister*, New York, August-September 1977.

CHAPTER 11
THE MASCULINE FACE OF SCIENCE

1. Most people are predominantly feminine or predominantly masculine, although most will display the opposite characteristics on occasion. Freud, among others, recognised that very few people have an exclusively feminine, or an exclusively masculine, psychology.

2. The feminine qualities that are, to some extent, positively valued in our society are almost entirely those related to child care. In general, most feminine attributes are traits having negative meaning in our society (such as weakness or dependence).

3. There are not very many women scientists of great renown, which Mozan, in *Woman in Science*, London, MIT Press, suggested was a result of history ignoring them, rather than that there never were any great women scientists. Given the prejudice that so many women in science have to face, a scarcity of women at the top is perhaps not so surprising. And it makes those who do get there – like Marie Curie, Rosalind Franklin or Dorothy Hodgkin – all the more remarkable.

4. Indeed, Bacon envisaged, in his book *New Atlantis*, a Utopia in which the state would be run by scientists, and its chief activity would be the development of technology. For a brief discussion of the rise of science during this period, see Rose, H. and Rose, S., *Science and Society*, Harmondsworth, Penguin, 1969, Chapter 1, and Easlea, B., *Liberation and the Aims of Science*, London Chatto & Windus for Sussex University Press, 1974.

5. At its most sinister, this aspect of control might result in attempts to change people's behaviour by surgical or drug intervention. Psychosurgery is one example, in which attempts are made to control people's behaviour by brain surgery. That these rarely achieve the desired ends is unimportant: the rationale given is terrifying. One example is quoted by Ackroyd, C. *et al.*, *The Technology of Political Control*, Harmondsworth, Penguin, 1977, p.276: a neurosurgeon justified surgical intervention by saying: 'Each young criminal incarcerated from twenty years to life costs the taxpayer perhaps $10,000. For roughly $6,000 we can provide medical treatment which will transform him [sic] into a well adjusted citizen.'

6. See, for example, Easlea, op. cit., and several articles in radical science publications, such as *Science for People; Radical Science Journal; Undercurrents*.

7. It could be argued the other way around – that because men were originally left out of the process of childbirth they developed, in compensation, a disregard for life, which led them to become ruthless and selfish, which in turn led them to start taking control over others. In other words, it could well be a chicken-and-egg question. I should emphasise here that I am not saying that all men are ruthless, or that all women are caring – I am talking about the values attached to things male and to things female.

8. The prevailing belief among investigators of the seventeenth to nineteenth centuries was that animals other than humans had no souls, and were, therefore, merely automata. One example is Francois Magendie, a neurologist of the early nineteenth century, whose *tour de force* was taking apart the brain of an awake rabbit while it thrashed

around in agony. Although animals are now usually anaesthetised prior to experiments, we still tend to accept this mechanistic approach to life.

9. Examples of the misuse of scientific knowledge are legion. Perhaps those who worked, like Einstein, on problems that ultimately gave humanity the power to destroy itself through nuclear weapons could foresee the horrors of Hiroshima and Nagasaki. Perhaps they did not fully comprehend, until it happened, just what power the atom could contain. Whatever their comprehension, atomic power is potentially the most awe-inspiring power that science has yet given us.

10. This is extremely unlikely with the present (capitalist) organisation of society.

11. Most people are by now familiar with examples of industrial pollution, some of which were not easily foreseen – such as the use of DDT as an insecticide: only after it had been used extensively was it realised that DDT becomes concentrated in the food chain, and kills many animals that prey on those animals that eat the sprayed crops. See, for example, Ward, B., and Dubos, R., *Only One Earth*, Harmondsworth, Penguin, 1972, pp.94–170.

12. Pointed out, for example, by Medawar, P., *The Art of the Soluble*, London, Methuen, 1967.

13. The imbalance is one that has been recognised lately, and that contributed to the rebellion of many young people against the technological society during the late 1960s and early 1970s – epitomised by Roszak's book, *The Making of a Counter Culture*, London, Faber, 1970.

14. Those who claim that IQ differences are largely a product of inheritance usually present their view in the guise of scientic truth, even though the data on which such views are based have little scientific basis. For a useful critique of this view, see Kamin, L., *Science, Politics and IQ*, Harmondsworth, Penguin, 1977.

15. It is worth remembering that IQ tests were originally developed precisely in order to discriminate between people. The original rationale was to develop a means to segregate educationally subnormal children from others in schools. These tests were later used to test would-be immigrants to the United States (see Kamin, op. cit.).

16. For example, a common view in the nineteenth century was that white males were the most evolutionarily advanced. Women, like men from other races, were thought to be less advanced, and consequently less intelligent.

17. The equation between male and person is exemplified by the attitudes of psychiatrists and psychologists studied by Broverman, I. K. *et al.*, 'Sex-role Stereotypes and Clinical Judgements of Mental Health', *Journal of Consulting and Clinical Psychology*, 34 (1970), pp.1–7.

18. Soren Kierkegaard; from *Stages on the Road of Life*, quoted in de

Beauvoir, Simone, *The Second Sex*, Harmondsworth, Penguin, 1976, p.175.

19. de Beauvoir, op. cit.

20. Sometimes literally. As atomic physics proceeds, smaller and smaller fundamental particles are discovered, and each time scientists begin to think of them as the 'building blocks of matter', until they can, in turn, be broken apart, like Chinese boxes.

21. Ivan Illich, in his book, *Medical Nemesis: The Expropriation of Health* (London, Calder & Boyars, 1975), points out that in an age of increasing medicalisation of life our culture has lost the ability to cope with, and to confront, death.

22. Not that some people don't try. For example, E. O. Wilson, in his *Sociobiology: A New Synthesis* (Harvard University Press, 1975), refers to the possibility of (eventually) subsuming all other areas of human concern (such as ethics and politics) under sociobiology – in other words, considering them as the product of evolution and genetics.

23. This is the reductionist approach, an attempt to understand the laws of nature by a fine analysis of the constituent parts. Less commonly, holistic techniques (that is, analysis of the properties of the whole, rather than its parts) are useful – for example, the study of the electrical waves of the brain, using the electroencephalogram (EEG).

24. For most intents and purposes, you can do this by using a very large number of people, randomly picked from the population, and randomly allocated to one of the experimental groups.

25. This is known as the 'double blind' method, for obvious reasons. The very fact that *not* using it can lead to spectacular distortions – the placebo group coming down with stranger symptoms than the group taking the drug, for example – shows that we know little about how drugs work at all.

26. The food we eat can have profound effects on us. It has recently been suggested that hyperactive children can be made less active by restricting their diets so that all artificial flavourings and colourings are omitted (one wonders what such additives are doing to all of us!). See Feingold, B. F., *Why Your Child is Hyperactive*, New York, Random House, 1975. This is perhaps a healthier approach to answering the 'problem' of hyperactive children than alternatives such as drug therapy.

27. The possible effects of radiation over a long period of time are in many ways difficult to estimate, since the deleterious effects of radiation are *cumulative*. That is, it is just as bad for you to be exposed to low doses of radiation over long periods of time as it is to be exposed to high doses for a short time. And scientists now recognise that the risks attached to radiation exposure are now greater than was once thought (see Morgan, K. Z., 'How Dangerous is Low-level Radiation?' *New Scientist*, 25.4.1979, pp.18–21).

28. Some checks on potential drugs are made before marketing – but only in the developed world. Most developing countries cannot afford or have not yet developed the extensive and elaborate checks that Europe and the United States carry out. But the failures of some of these checks indicate that even we aren't careful enough over checking – see, for example, Weiss, Kay, 'Vaginal Cancer: an Iatrogenic Disease', in *International Journal of Health Services*, 5 (2) (1975), pp.235–52; and Seaman, B. and Seaman, G., *Women and the Crisis in Sex Hormones*, New York, Rawson, 1977. The latter authors also suggest that there might be an increased likelihood of cancer of the scrotum in boys born to mothers who were given DES. Perhaps the sickest part of the whole DES story is that it is still prescribed in both the United States and Britain – as a 'morning-after' pill.

Name Index

Subject Index

Virago

If you would like to know more about Virago books, write to us at 5 Wardour Street, London W1V 3HE for a full catalogue.

Please send a stamped addressed envelope

VIRAGO ADVISORY GROUP

Andrea Adam · Carol Adams · Sally Alexander
Anita Bennett · Liz Calder · Bea Campbell
Angela Carter · Mary Chamberlain · Deirdre Clark
Anna Coote · Jane Cousins · Anna Davin
Rosalind Delmar · Zoë Fairbairns · Carolyn Faulder
Germaine Greer · Jane Gregory · Christine Jackson
Suzanne Lowry · Jean McCrindle · Nancy Meiselas (USA)
Mandy Merck · Cathy Porter · Elaine Showalter (USA)
Spare Rib Collective · Mary Stott · Anne Summers (Australia)
Rosalie Swedlin · Margaret Walters · Alison Weir
Elizabeth Wilson · Women and Education
Barbara Wynn

Book Tokens

Give them
the pleasure of choosing

Book Tokens can be bought
and exchanged at most
bookshops.

OTHER BOOKS OF INTEREST

WOMEN IN WESTERN POLITICAL THOUGHT
Susan Moller Okin

Through a detailed study of four political philosophers whose
ideas on the rights of man have profoundly shaped our own –
Plato, Aristotle, Rousseau and Mill – this book looks at why,
despite gaining formal rights of citizenship, women still do not
have full equality with men.

'An excellent book. . . . her language is calm, clear, simple and
strong, her feminist perspective a line of thought that guides her
steadily along' – Vivian Gornick, *Washington Post*

HALF THE SKY
An Introduction to Women's Studies
eds. The Bristol Women's Studies Group

An anthology of famous and lesser known text from British,
American and European sources providing a key interdisciplinary
reader on women's lives and experiences.

'A wide ranging text which will interest the general reader as well
as participants in women's studies courses. . . . The wealth of
source materials quoted make accessible a great deal of women's
writing' – *Tribune*

FEMALE SEXUALITY
New Psychoanalytic Views
ed. Janine Chasseguet-Smirgel

A remarkable addition to the study of female sexual identity,
which, while acknowledging the power of culture and society in
shaping our lives, explores the profound influence of instincts and
the unconscious on our experiences of sex and love.

'Lucid and not too loaded with scientific terms . . . Some valuable
insights.' – *Publisher's Weekly*
(1980)

WOMEN IN SOCIETY
Interdisciplinary Essays in Women's Studies
Cambridge Women's Studies Group

Combining original research with critical examination of existing arguments, this authoritative book, based on the course of the same name at Cambridge University, offers new perspectives on themes central to the understanding of women's place in society. (1980)

OF WOMAN BORN
Motherhood as Experience and Institution
Adrienne Rich

With a blend of personal memoir, history and anthropology, Adrienne Rich, poet, critic and mother of three sons gives in this famous book a poignant and illuminating account of motherhood in a patriarchal society.

'Rich's chapter on the relationship between mothers and daughters would alone be enough to cherish her book.' – Jill Tweedie

SEXUAL POLITICS
Kate Millett

Devastatingly witty and brilliantly researched, this passionately argued work of literary and cultural analysis is one of the most famous critical works of our times.

'A book which must take its place at a single leap in the ranks of the best modern polemical literature.' – Michael Foot, *Evening Standard*

NOT IN GOD'S IMAGE
Women in History
eds. Julia O'Faolain & Lauro Martines

Through the words of contemporary writers this invaluable source book explores the lives women have led throughout European history, illuminating the position of women over the centuries.

'It adds a whole new dimension to the study of history, from antiquity to the present.' – *New Society*

PATRIARCHAL ATTITUDES
Eva Figes

With Germaine Greer's *The Female Eunuch* and Kate Millett's *Sexual Politics*, *Patriarchal Attitudes* was a book which changed the consciousness of a generation, and has established itself as an outstanding contribution to the debate on women's position in society.

'The only one whose work can be set beside John Stuart Mill's celebrated review of the subject and not seem shoddy or selfserving.' – Gore Vidal, *The New York Review of Books*

ON LIES, SECRETS AND SILENCE
Selected Prose 1966–1978
Adrienne Rich

An important collection by one of America's finest poets and feminist theorists, including essays on literature and politics, women's history and culture, ethics and aesthetics, racism, sexism and class oppression.

'Adrienne Rich's prose moves with force, clarity, energy, it soothes with a poet's grace and elegance.' – Ellen Moers

WOMAN AND LABOUR
Olive Schreiner

This famous book, which analyses women's economic position in society, is one of the most profound and eloquent pleas for a new relationship between the sexes, its impact today as great as when it was first published in 1911.

'Strong, large in scope, generous, bold, Olive's subject is humanity, and a woman as she can contribute to the development of humanity.' – Doris Lessing, *Guardian*